S0-BBT-935

CV: CARRIER AVIATION

CV: CARRIER AVIATION

By Peter Garrison   Photographs by George Hall

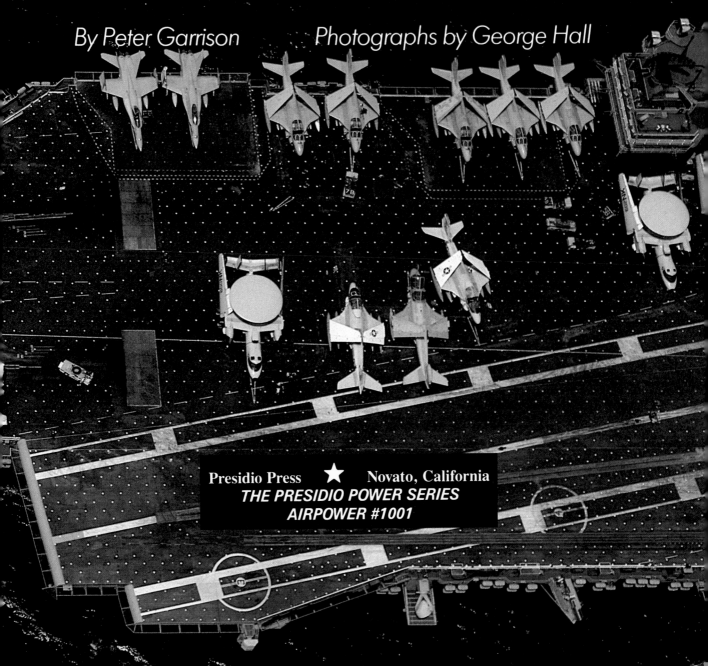

Presidio Press ★ Novato, California
THE PRESIDIO POWER SERIES
AIRPOWER #1001

Copyright© 1980 by Presidio Press
Revised edition, 1984, 1987

Published by Presidio Press
31 Pamaron Way, Novato, CA 94949

All rights reserved. No part of this book may be reproduced or utilized in any form or by any means, electronic or mechanical, including photocopying, recording or by any information storage and retrieval systems, without permission in writing from the Publisher. Inquiries should be addressed to Presidio Press, 31 Pamaron Way, Novato, CA 94949.

Library of Congress Cataloging-in-Publication Data

Garrison, Peter.
　　CV, carrier aviation.

　　(Presidio power series. Airpower; 1001)
　　1. Aircraft carriers--United States.　2. United States.　Navy--Sea life.　I. Hall, George (George N.) II. Title.　III. Series.
V874.3.G37　1987 359.3'255'0973　　　87-3038
ISBN　0-89141-299-9　(pbk.)

All photographs copyright George Hall, with the following exceptions:
　　P 68:　LT Evan Edwards USN
　　P 53 (top):　CDR Dave Erickson USN
　　P 88:　Goodyear Aerospace
　　P 63:　Grumman Aerospace
　　P 2-3, 35, 58, 62, 84-85:　Robert L. Lawson/Tailhook Photo Service
　　P 48, 52, 53 (bottom), 75, 76, 79, 86, 90, 92:　US Navy

Printed in the United States of America

Contents

Acknowledgments

The United States Navy was cooperative in arranging visits to several carriers for the author and photographer. We thank Vice Adm. Robert P. Coogan, COMNAVAIRPAC, and Vice Adm. George E. R. Kinnear II, COMNAVAIRLANT, and the officers and crews of *Constellation*, *Kitty Hawk* and *Nimitz* for their hospitality and assistance. Special thanks to Lt. Comdr. Dave Erickson, LSO of VF 111, for the exceptional efforts he made to provide us with information, photography, and insight into carrier training and operations; to Robert Lawson, editor of *The Hook*, for his editorial comments and for access to his immense photographic files; to Rear Adm. Paul A. Peck, USN (Ret.), formerly Commander Carrier Group 3, for his painstaking and very helpful review of the text for accuracy, and to Comdr. Tom Jurkowsky and Senior Chief Fred Larson at Public Affairs COMNAVAIRPAC, NAS North Island. Without the help of these people, and many others, this book would not have been possible; but the author retains for himself credit for any errors which may have slipped past his guides and mentors.

THE CASTLE IN THE SEA

As YOU board the helicopter for the ride out to the carrier, you are given a life vest and a "cranial." The cranial, a kind of segmented helmet, fits snugly over your head, with ear protectors that grab your ears. Within the cranial, all sound is muted. The eighteen passengers on board the helicopter are made up of enlisted men, pilots, a couple of ship's officers, and the chief of staff, a seasoned captain with a boyish but weary face.

It is lonely inside the cranial. You can't talk to anyone. Like travellers of different nationalities in a bus, the passengers sit solitary and impassive. Some lean back against the vibrating, unpadded wall of the cabin; others hunch forward, elbows on knees. A few appear to sleep, but it is more likely that by closing their eyes they are merely making the final step in the withdrawal from fellowship, which the cranial, the noise, and the discomfort compel. The helicopter is shabby. No one smiles. The flicker of rotor blades interrupts the diffuse sealight. Among the lolling heads one man looks frightened; his thumb is pressed against his teeth, his eyes are wide, his skin moist. On the floor, at the ends of randomly angled legs, are neat rows of brilliant black shoes.

Only the crew chief is exempt from the isolation. He moves about, crouching, one hand on his boom mike, one stretched upward to a handhold like a gibbon's, talking to the two pilots. But no words are heard, only a rattling, a whining and shaking, the noise of something metallic clunking overhead.

Enterprise (CVN-65), one of four nuclear carriers. Previous page: Enterprise entering San Francisco Bay.

2

The horizon cannot be seen. One distinguishes between the water and the sky only from force of habit. The travellers sit immobile, suspended between being and nothingness, scarcely blinking. The crew chief looks with concern from the passengers to his watch. He makes a note in a green leatherette booklet. His helmet has yellow lightning bolts and arrows on a green field. No one seems to be breathing.

Suddenly, in the frame of the doorway, interrupting the gray expanse of ocean, appears the outermost rampart of a floating city—dark, stationary, and strange.

WHETHER THE carrier is huge or not depends on your point of view. It is certainly a big ship; some of the fourteen carriers now in the service of the United States Navy are the largest warships ever built. They are smaller than supertankers, but larger than ocean liners. Their largeness is two-dimensional, a flat surface of four acres or so, like a parking lot, black, striped with white and yellow lines, and coarse-surfaced.

From certain vantage points the deck resembles the roof of a gigantic skyscraper; at its edge, everything seems to stop at once. Beyond it the ocean flows past like a far-off river. The illusion of the platform remaining stationary while the ocean itself streams by is more persuasive than you would think without seeing it. From the middle of the deck only a distant portion of the sea, the few degrees below the horizon, is visible. It seems to move past in fixed rank, devoid of the spreading, fanlike motion of perspectives possessing a foreground, like the countryside one sees from a train or, for that matter, the sea one sees from any ship that

is not as flat as a table. Without a fast-moving foreground to explain its slowly flowing background, the ocean becomes a mere ribbon of water: a river. And so the ship, though it pitches and heaves now and then with a ponderous slowness, gives predominantly the impression of standing still.

This sense of immobility is complete when one is below decks. In the hangar deck, with its cavernous oval openings at the aircraft elevators, one feels exactly as if standing in a large warehouse whose loading ramp looks out on the Mississippi or the Amazon. Still farther below, where there is no view of

4

Left: *F-18 and A-7 attack jets are spotted forward on four-acre flight deck of* Constellation *(CV-64). Above: F-14 Tomcats, their wings stowed, are arrayed aft.*

the ocean, it is like being in some enormous and puzzling subterranean structure, the multiple basements of a skyscraper, perhaps. Miles of passageways and hundreds of steep ladders thread their way from compartment to compartment. Most of the active carriers measure over 1,000 feet from bow to stern, 250 feet in beam, and over 200 feet above the waterline. Their dimensions are similar to those of a large building; in fact, a carrier resembles an 80-story building lowered onto its side, equipped with engines and set afloat.

In this structure some 5,000 men live. The officers live in small, spartan spaces; only the commanding officer, and the admiral whose flagship the carrier is, are entitled to a small ration of luxury. They drink iced tea from stemware. At the other extreme are the several thousand enlisted men who are the laborers in this little city and who sleep in bunks called "racks," stacked four or five deep like shallow shelves, between the deck and the seven-foot ceiling. In these, a man awakening with a lurch from a dream of being buried alive will find, in the faint red light that glows both day and night, that he is not dreaming at all.

5

Within most of the ship there is neither day nor night. The sleeping spaces and passageways are lit at all hours. The engines never idle. No hour is preferred to any other for work or rest. Missions can be launched and recovered around the clock, but the height of activity usually runs from early afternoon until the small hours of the morning. When a recovery is complete the men can doze for an hour or two. The rotation of the earth for them is insignificant; they rise and retire to the wheeling not of stars and planets, but of squadrons of jets. In the small appendices of the main passageways, where a little-used ladder leads to an unfrequented space, two or three young men may be found curled up on the steel floor, resembling the mysterious sleepers that one encounters snuggled against a wall in the poor quarters of an Arab city.

The ship's company is distinct from the smaller complement of men who serve the airplanes and the pilots. Aboard a carrier, roles are clearly defined. On the deck, the color of the jacket reveals the function of the man, and often his position in a hierarchy of jobs.

The blue shirts are chock and chain men who carry and position the heavy blocks that are placed beneath the wheels of planes to prevent their rolling, and the chains that secure the planes to the deck. When an airplane is moving to its parking place, two men in blue jackets, hunched over and hefting chocks, walk alongside its main wheels.

A stencilled **E** on the blue jacket means a deck-edge elevator hand. **T** means tractor driver, the pilot of one of the boxy tugs that tow the heavy airplanes around the deck. **T** has a certain prestige, handles his own machine, and gets to sit down; so it is desirable to be a **T**.

More desirable, however, and attainable if one possesses the aptitude and experience, is the yellow shirt. Yellow is a taxi director: he has absolute authority over the movements of the airplanes on the deck.

Flight-deck crews wear jerseys and life jackets that are color-coded to their jobs.

6

Deck supervisors wear two-way "mouse" radios to communicate with the Air Boss, the hangar deck, and each other.

specific airplanes. Symbology varies from ship to ship, but typically, diagonal grooves in an airplane counter mean it is out of service; one purple nut atop the counter means it needs fuel; two mean it needs defueling; and so on. Color codes identify the squadrons.

The Handler communicates with the directors on deck and in the hangar deck through self-contained, two-way FM radio-headphones. Each man wears a set—called a "mouse" because of Mickey Mouse ears, listens for instructions, and hears, every few seconds, with an eerie and metronomic regularity, a beep or "confidence tone" that assures him, like the dial tone on a telephone, that his receiver is still working.

Nothing budges without his directions. His walkers guard the wingtips and tails, gesticulating and shouting as the big airplanes roll slowly, clearing each other and the superstructure of the ship by inches. Sometimes they fail to clear: fatigue takes its toll.

Rearranging the airplanes on deck is reminiscent of games that consist of sliding tiles in a frame, where it is necessary to move every tile several times in order to place a certain one in a certain position. The ship's puzzle is rehearsed, in miniature, on a table at the bottom of the island, at flight deck level. The flight deck itself cannot be seen, but on models of the deck and the hangar deck below, counters of the proper shape and scale are experimentally moved about, and the real airplanes are then placed accordingly. The Handler or "mangler," as the officer in charge of this airfield in miniature is called, identifies the features of

Movement of all aircraft on deck is controlled from the Handler's table.

7

The airplanes are fueled by teams in purple shirts, fuel arriving under pressure from deck-edge pumps through hoses as big as fire hoses that connect to a single point somewhere on the airplane and force-feed fuel into all its tanks at once, pumping thousands of gallons in a few minutes.

Red shirts handle ordnance, load the bombs onto the racks, the rockets onto their pylons, and the ammunition into the guns; red shirts with black stripes drive the fire and crash trucks. Green shirts—hookup men—ready the airplanes for launch, securing them to the catapults. Green shirts with black stripes are the airwing men, who perform all maintenance on the aircraft. Brown shirts—plane captains—are responsible for the care and cleanliness of each airplane; they also ride the brakes when the aircraft is moved on the deck. White shirts with checkered helmets, called "checkers," inspect the airplanes coming up to the catapults for launch and give the thumbs-up signal if they're okay. A lone, princely figure

on the sidelines is dressed in a silver fireproof suit. He has one job only: if an airplane crashes, he must run to the wreckage and rescue the crewmen from the flames.

Each man attends "P" school—a three-week orientation that includes flight-deck theory and first aid. After that his job is passed on to him by example and word of mouth. The novice takes his chances; he learns by working alongside others who have been doing the work for weeks or months before him. The deck is a hard school: they learn fast here and soon are shouldering responsibilities beyond their years. They are shaped by their positions. Positions change hands, but each has its own protocol, strict and exclusive. The choreographies of the catapult crews are as uniquely their own as the mating dance of certain birds. The prerogatives of the directors are as exclusive as those of the pilots. Like a Chinese puzzle consisting of dozens of intricately shaped wooden pieces which, when correctly assembled, hang together as a smooth and faultless cube or sphere, the carrier is a functional unit constructed of dozens of fraternities, disparate, xenophobic, and proud.

The flight deck is oblong in shape, with a tapering bow and stern. It is smooth and featureless. Most of the equipment on it can be retracted and stowed. Standing on the deck, one is only dimly aware of the angled runway, the positions of the four catapults in convergent pairs, bow and waist, the elevators, the barrier net, or the jet blast deflectors; the dominating feature is the

Fire-fighters in silver suits stand by during launches and recoveries.

8

Island on Nimitz, with Primary Flight Control (Pri-Fly) — domain of the Air Boss —visible beneath the bridge.

superstructure called the "island." Located on the starboard side, approximately amidships, it resembles different things on different carriers. The *Enterprise's* original island resembled a modern office building, sparsely rectilinear and crowned by a conical structure that suggested a rotating restaurant. Its radars were incorporated in the flat sides of the tower. No other carrier was built after that pattern, however, and the *Enterprise* was recently overhauled to modify the novel island.

Most islands resemble the fortresslike structures on old cruisers and battleships, but inverted so that their greatest bulk appears not at the base, but somewhere near the top. Towering above the island is a mast sprouting a demonic foliage of round and square radar reflectors, antennas, catwalks, crossbeams, booms, cables, stays, and festoons of ropes and pennants: a surrealistic tree like those in the hell paintings of Bosch. The radar antennas swing restlessly, scanning, searching, tracking and guiding, ablaze with invisible radiations that could supposedly vaporize an unwary seagull. Considerably longer at its base than it is wide, the rest of the island is a puzzling assembly of cylinders, rectangles, slopes of thirty and forty-five degrees, cantilevered balconies and mezzanines, ladders, pipes, hatchways, and rows of opaque greenish windows like the optical sensors of some sinister robot.

The principal row of windows belongs to the bridge on which an immemorial captain

9

Captain Tom Mercer, skipper of Carl Vinson (CVN-70), in his leather swivel chair on the bridge.

swivels in an elevated barber's chair commanding a view of the deck and issues orders to the helmsman who spins a brass wheel. The officer of the deck sets the course and speed of the ship based on recommendations from the navigator. For launches and recoveries, the ship must steer a course into the wind that will create a headwind of twenty-five to thirty knots over the angled deck. In calm weather, this means full engines in any direction and settling for a slight right crosswind on the deck. But usually there is a wind at sea, and the officer of the deck calculates the heading and speed relative to the wind in order to

On the bridge of Constellation (CV-64).

obtain the desired wind over the deck. On the basis of readings of the apparent wind on the deck and the ship's heading and speed, he can solve for the true wind, which is the speed and direction of the wind over the sea independent of the wind created by the ship's movement. He then reverses the procedure and solves for a ship's heading and speed which, combined with the true wind, will give the required relative wind. When the carrier reverses direction—it takes about ten minutes, generally, to complete a 180-degree turn—there is a moment when suddenly your wandering attention falls upon the wind, and you realize that it has disappeared. The steady thirty-knot blow seems to have been replaced by a calm. Yet the water is still agitated and foam blows from the crest of waves. What has happened is that the ship, in changing course through downwind heading, has subtracted its own velocity from the wind's and created the illusion of calm weather. Smoke billows indecisively from the funnels on the island, and the signal pennants hang limply.

The airplanes huddle in dense rows on the flight deck and the hangar deck. If anything makes the carrier look small, it is the airplanes, which are far larger than you might expect, today's fighters being about the size of the medium bombers of World War Two. To make them more compact, their wings fold for storage. The folding mechanisms are hydraulic. One set of rams pulls hardened steel pins out of the shear fittings; another cylinder drives the wing upward or backward. The cycle is automatic, occurring at the press of a button, and the airplanes begin folding their wings the moment they roll away from the arresting

10

Above: *Wing-folding mechanism of the A-6.* Right: *Each type of carrier aircraft folds its wings differently.*

gear on recovery, unfurling them again only as they roll onto the catapults for launch.

Each type has its own way of folding. The swing-wing F-14 has a position of a wing sweep, steeper than the steepest used in flight, for deck storage. The wing roots of the F-14 angle upward and the outer panels, when swept, seem to droop back downward. The thickness of the root, the slenderness of the outer wing, and that hunched shape are so strikingly like those of a bird's wing that seeing the parked F-14 head-on you cannot help feeling that you are looking at a hawk in that moment of flight when, wings partially folded, it swoops downward to its prey.

The S-3 Vikings fold their tall vertical tails over to one side and their long slender wings up vertically, the right wing slightly forward, the left slightly aft, so that they cross in the middle. The A-6's wings fold upward at midspan and meet over the center of the airplane. But the E-2 Hawkeye, its huge rotating radar saucer making such a folding arrangement impossible, uses a system similar to that of many light sport

11

airplanes, one in a way more elegant and birdlike than any other: the wings pivot on a rotating hinge, leading edge downward, and swing backward at the same time, so that they end drawn up flat along the flanks of the fuselage, exactly like the wings of a perching bird.

Most astounding are the contortions of the SH-3 Sea King helicopters which, adding the complexity of hydraulic folding to what is already a mechanical Gordian Knot, automatically fold their five rotor blades like complicated jackknives, and curl against the fuselage the end of their angled tail booms, antitorque rotor and all.

Wings and tails folded, the airplanes fit into remarkably little space; the big carriers can handle more than a hundred of them, of which sixty to eighty may be on the flight deck during operations, the rest on the hangar deck. They are shuttled back and forth between the flight and hangar decks on four elevators located along the edge of the decks. Within the hangar deck the airplanes park in incredibly tight rows, separated only by a hand's breadth. It is here that the airplanes are stored and maintained. Through the night, in pools of light falling from a high ceiling crisscrossed with steel beams and pipes, technicians huddle in the cockpits of the airplanes and peer through their removable access doors at knots of metallic viscera. The hangar deck has its own jet engine repair facility. Electronic equipment, most of it packaged in removable modules, is serviced at a computerized testing facility that automatically analyzes a problem, printing out its diagnosis on a cathode ray tube.

Airplanes being moved about prior to launch are hauled by small tractors called "tugs" or "mules," or may be pushed by hand; after recovery they move to temporary parking under their own power. Getting a twenty-ton airplane rolling, even slowly, can take a powerful burst of thrust; sometimes men, or even other airplanes, are blown overboard by the unexpected, furnace-like, fuel-smelling blast of a swinging tailpipe. But this is only one of the hazards for the men on the deck. A man can be fished back out of the sea; but when, on occasion, a man passes too close to the intake of an idling engine, he is seized and devoured.

Left: *Sea King helicopters aboard* Constellation, *with main rotors and tails folded.* Right: *Joggers dodge parked jets in cavernous hangar deck.*

12

The men never tire of talking about the dangers of the deck. They have heard of arresting cables breaking and whipping right through a man, slicing him in half; they have seen the films of the catastrophes on the *Forrestal* and *Enterprise* where a runaway rocket skittered forward from the fantail igniting ordnance, airplanes, and men; they have seen the films of airplanes hitting the rearmost edge of the flight deck, breaking apart, flinging flaming fragments down the deck; they know of the dangers of the old jet fuel, JP-4, which has been replaced by the far less hazardous JP-5, but they also know that even JP-5 will burn furiously once ignited. Even though the deck is equipped with an emergency washdown system in case of a "conflagration," they know that fire is still to be feared.

They know the dangers of fatigue. These ship's boys do not sleep peacefully in the crow's nest, rocked by the tempest; they work long hours, beyond exhaustion, in the cold wind, in rain, and in the stifling jet blast that envelops a man and deprives him of

oxygen until panic wells up in his lungs. They can tell you about the shudder of fright that runs through you, making your knees rubbery, when, leaning against a jet on an idle deck you close your eyes for a second to rest and on opening them find the deck full of planes preparing for a launch.

Danger is familiar. They come to joke about being sucked into the jet engines — which does in fact happen about once a year—and almost everyone has, at one time or another, come close, in his opinion, to being blown over the side. In addition to all the specific, detailed dangers they can enumerate, there is the encompassing, per-

Flight-deck workdays often exceed twelve hours; fatigue is one of the dangers of the deck.

14

vading one: that of standing in darkness on a target at which a fifteen-ton airplane aims at 150 knots, with only a fallible man at its controls—a danger bizarre, incalculable, and entirely routine.

Intake of the A-7 has a particularly sinister reputation.

15

CUTAWAY

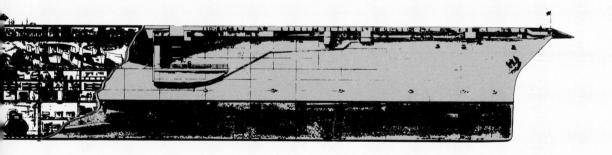

THERE ARE fourteen carriers in the U.S. Fleet. The *Nimitz, Enterprise, Dwight D. Eisenhower,* and *Carl Vinson* are powered by nuclear reactors; the rest are powered by oil-fired boilers. The conventional carriers are the *John F. Kennedy, America, Constellation, Kitty Hawk,* which are, with the nuclear ships, the largest and most modern; and the *Independence, Ranger, Saratoga, Forrestal, Coral Sea,* and *Midway.* The last two were built just after World War Two; they alone are excluded from the category of "supercarriers." Another carrier, *Lexington,* is used as a training ship only.

The *Forrestal, Kitty Hawk,* and *Kennedy* class ships are just over 1,000 feet long, with a 35-foot draft, 130-foot wide hull, a maximum flight deck width of over 250 feet, and a maximum displacement of almost 80,000 tons. These dimensions are not much inferior to those of the largest carriers, the *Nimitz, Enterprise,* and *Carl Vinson,* which are about 60 feet longer than the *Forrestal,* equally wide, and displace an additional 10,000 tons.

All the supercarriers have a rated power of between 200,000 and 280,000 shaft horsepower driving four screws, and are capable of speeds of thirty to thirty-five knots. Stories of the phenomenal speed of the nuclear ships are certainly exaggerated under the inspiration of their mysterious powerplants. Commissioned in 1961, the *Enterprise* alone of all the supercarriers was built with the distinctive pineapple-in-a-box island with its billboard-like fixed radar arrays; now that this has been replaced by a more conventional island, the remaining differences among carriers are comparatively subtle. The *Forrestal* was redesigned in the course of construction (1952–54) to incor-

porate the angled deck pioneered by the British, which seems in retrospect such an obvious advantage, but which was so long in coming. The *Forrestal* also incorporated steam catapults, another English development, in place of the earlier hydraulic ones. The general arrangement of the *Forrestal's* deck differs from that of later designs, however, in having one rather than two elevators forward of the island.

The island is usually about 50 feet tall, 20 feet wide at the base, and about 100 to 120 feet long. It cantilevers outward at the top, clear of the moving airplanes. At its base, there are entryways on both seaward and deck sides. Those facing the deck are usually closed, both to protect the island from deck fire and to keep noise, debris, and jet exhaust from filling its spaces. Along the front of the island, 30 feet above the deck, are the rows of slanting green windows of the bridge; on its port side, toward the aft of the island, is the projecting gazebo of the Primary Air Controller, and below that a TV camera turret. On the broad roof beneath the radar antennas is an observation deck called "Vulture's Row."

On the nonnuclear ships the aft half of the island contains the funnels. The exhaust is very clean, so thorough is the burning of the fuel. A periscope in the boiler room allows one to look up at the blue sky through the stacks to assess the quality of burning; normally no clouding of the air is visible at all.

The ship is layered from stem to stern in decks; the main deck is the hangar deck, a vast enclosed space over thirty feet deep. It is much narrower than the flight deck four levels above it, being within the hull and lacking the additional breadth of the flight deck's cantilevered overhangs. Narrow

19

Above: Coral Sea (CV-43), a World War II design, was modified in the 1950s to incorporate a twelve-degree angled deck. Left: A-6 on forward elevator (EL-1) of Constellation.

passages on the sides of the hangar deck connect the lower decks with the flight deck and the island, which itself is cantilevered out over the water. The four elevators, which are used to carry airplanes (often two at a time) between the flight deck and the hangar deck, and which each can lift 130,000 pounds, are movable portions of the flight deck overhangs which drop down to oval openings sixty feet long and twenty-five high in the sides of the hangar deck.

The hangar does not occupy the full length of the hull; it is about 600 feet long and centered approximately in the middle of the carrier. The forward part of the hull is given over to other spaces, including living

20

spaces which, on occasion, might be situated right next to a catapult and directly beneath the flight deck. During night operations one can hardly talk in a stateroom next to catapult machinery, but a tired pilot sleeps undisturbed.

The maze of compartments beneath the hangar deck is threaded with fore-and-aft and transverse passageways. The passageways are narrow; if two men meet going in opposite directions, one must turn sideways to let the other pass. Every thirty feet or so is a bulkhead or frame with an oval opening called a "kneeknocker" through which one steps. Walking along a passageway is a rhythmical dance: stride, stride, stride, STEP, stride, stride, stride, STEP . . . .On most carriers all the openings in the bulkheads are identical and in perfect alignment. The lighting is flat. Seeing another person approach in the distance is like seeing yourself approach in a mirror down a succession of arches endlessly reflected.

Each compartment is numbered with a code that fixes its location; but the passageways are not continuous, nor is their grid symmetrical. Some sidestep; some come to a dead end; others continue but without an outlet upward or downward. There are ladders only at certain points; they connect only certain groups of decks; they are often hidden in unmarked byways. And so, in spite of the rational system of compartment addresses, it is almost impossible for a newcomer to find his way around. The carrier is like some Mesopotamian medina where, after wandering in bewilderment for hours, you must ask some passing boy to show you the way out, which invariably proves to have been quite close to where you stood.

21

The average age in the carrier crew is less than twenty.

The ladders connecting the decks are extremely steep; descending them it is easy to miss a step altogether. Some have handrails; the ones between the bridge and the Captain's stateroom have bannisters elaborately decorated with varnished rope fancywork; most have for a handhold an unceremonious grey-oily chain.

Counting up from the main deck the levels are numbered 01, 02, 03, 04, etc. (The flight deck is usually on 04 level.) Below the main deck, they are numbered 1, 2, 3, on down. The waterline is around 2 or 3, and there the openings in the deck are equipped with watertight hatches. When battle stations are called, the watertight bulkheads are closed; men who are below the waterline stay there. On one of the Japanese carriers sunk in the battle of Midway, hundreds of men were left in the engineering spaces when the ship was abandoned. When the ship failed to sink promptly, a crew returned and encountered some of the engineers, who had finally come topside to find out why they had heard nothing from the bridge for several hours. What was said on that occasion has not been recorded.

Most of the conventional carriers use a propulsion system incorporating eight or twelve boilers to produce high-pressure steam. The steam drives eight turbines connected to the final drive shafts through reduction gear devices each about the size of a small bedroom. The driveshafts, about twenty inches in diameter, pass through the boiler spaces, marked with orange stripes of warning. The powerplants are located amidships in spaces of incomprehensible shape, gorged with machinery, many decks away from daylight. Huge ducts blow cool outside air into the machinery spaces, which

22

nevertheless are seldom cool, having only isolated sluices and pools of coolness. Through the floors of metal gridwork, the inside of the double hull can be seen, rusty and scabrous, under an uninterrupted sheet of flowing water. Everything is covered with a slick emulsion of water and oil. The footing on the narrow catwalks, threading their way among the hot pipes and tanks and the valves, is precarious. The atmosphere is dense with steam, and the temperature sometimes passes 110 degrees.

The men and boys who work in those moist hot caverns go by the nickname of "snipes"; they are joked about by deck-hands as a strange breed, nocturnal, pale, forgetful of the sun. Sometimes, however, a snipe climbs the twenty levels of steep ladders to Vulture's Row, just below the big radar arrays, to watch the deck operations and take the air. Then he stands in his T-shirt in an icy wind, returning to the world some of the surplus of heat he has absorbed below.

In place of oil-fired burners, the nuclear carriers use nuclear piles or reactors to produce steam. The *Enterprise* has eight of these, the *Nimitz* only two. Their power rating is determined not by the enormous ability of the reactors to produce heat, but by the ability of the peripheral equipment — turbines, gearboxes, pressure vessels — to handle the power. Thus the power rating of the nuclear carriers is little different from that of their conventionally powered counterparts.

Nuclear reactors differ from conventional steam systems in that an oil burner must carry a fuel supply that requires replenishment relatively frequently, and in that the fuel must be fed to the burners at a certain rate to provide the proper conditions for combustion. A shipboard nuclear reactor consists essentially of a container full of water in which is immersed a quantity of uranium. When enough uranium of proper chemical purity and isotopic composition is assembled in one place, it gets hot automatically. The problem is not, as with oil, to get it hot enough, but to prevent it from spontaneously getting too hot; this is done by inserting "control rods" of a different material into spaces in the mass of uranium. The control rods modulate the self-heating of the uranium pile.

Being radioactive, the uranium pile must be surrounded with heavy shielding to protect the crew. The omission of fuel oil compensates for the weight of this shielding. A conventional carrier carries about 2.2 million gallons of fuel for its own boilers (in addition to about 2 million gallons of a different type fuel for its aircraft); the uranium fuel for the reactors weighs a tiny fraction of those 7,300 tons. It has the further advantage of lasting much longer; the 1971 refuelling of the *Enterprise* — an operation which cost almost $100,000,000 — was expected to last thirteen to fifteen years, with the ship cruising over 60,000 miles a year for a direct cost of about $133 per mile. Conventional carriers at normal cruising speed burn about $250 of fuel per mile; but the figure is not really comparable, because it does not take fuel storage and tankering costs into account.

Carriers are in any case obliged frequently to replenish their stores of aviation fuel, ammunition, and supplies for their huge crews. Operating off the coast of Vietnam, the carriers were unthreatened and could take on supplies at their convenience; some carriers regularly replenished every three days, rendezvousing with supply ships and taking on stores through elaborate hookups of hoses and cables. It is obvious that with storage facilities for two million gallons of aviation fuel, and fifty or sixty active airplanes each burning two or three thousand gallons per mission and flying twice daily, a carrier cannot go unreplenished for long. Nuclear carriers are at an advantage in operations far from supply bases; they can carry two-thirds more supplies and hence replenish less often.

The carriers generate many stupendous statistics. The most populous, the *Nimitz* and *Carl Vinson*, have a full complement of 6,300 souls; 5,000 to 5,500 is more typical of the other carriers. Each of these men presumably eats three meals a day, though some miss their meals and eat junk food at the "gedunk" (long "e") instead. More than fifty berthing compartments serve to store enlisted men while they are asleep. The complement of planes is often over 100. The four screws that drive the big ships can be 21 feet in diameter; each rudder can weigh 45 tons; each anchor 30 tons, each link in the anchor chain, 360 pounds. And so on.

The statistics convey a sense of bigness remote from everyday experience. Nor does it matter to the crewman on a carrier what the anchor links weigh, or how heavy the rudder is. What he is aware of, instead, is the urban, impersonal quality of the ship. Everyone comments upon it; carrier duty is not popular among those who, not being part of the Air Wing, could be and have perhaps been on other kinds of ships. A destroyer, they say, is like a small farm town: not many places to go, but you know everyone you meet. Any ship is that way—except a carrier. A destroyer may have a crew of 250 to 400 men; a cruiser 400; but the 5,000 or 6,000 men who live on a carrier are numbered in proportion to the cube of the ship's linear dimensions. The men are seemingly added to fill three-dimensional space to a certain density. Since the space is broken into hundreds of tiny compartments, one never sees a great many men at one time, unless on the flight deck for a FOD (foreign object damage) walkdown or barrier rig drill. The population is ordinarily betrayed not by the sight of vast crowds, but by the constant encounters with strangers. As in a city, one may never get to know one's next door neighbor; after all, he works and eats in a different place.

Enlisted men's "racks."

24

Compartmentation is thought, in animal populations, to reduce the competitive pressure of overcrowding by making most of the population invisible to each individual most of the time. Overcrowded mice and rats do not fight so long as they can remain hidden from one another, however small the space available to each individual. Humans might plausibly behave in the same way. If claustrophobia engulfs the inhabitant of the carrier, it is not brought on by his fellow sailors, but by the walls and ceilings that are always close by, indifferent, unornamented, and monochromatic.

As with any community, not all of the members hold jobs directly related to the community's principal mission. In every coal town someone is a druggist, someone runs a movie theater, someone mends shoes. So on the ship: many of the crewmen are in service occupations, laundry, food preparation, cleaning, maintenance, entertainment, religion, medicine. There is a hospital with an operating room, a dentist's office, a barber shop, two television stations (which produce, among other things, the daily briefings for the flight crews), a radio station, a newspaper, a library, a chapel, a weight room, and a hobby shop. Down in the tunnels of the ship's interior, hidden populations go about daily routines remote from the dramatic and well-publicized activities of the flight deck. One glimpses, through open doorways, the personal vistas, the individual dreams, diversions, obsessions and affections. Leaning against a stack of radio equipment pulled out of airplanes for repair is a guitar. On a wall above a work table is a picture of Linda Ronstadt. A model sailboat, dusty and battered, lies heeled over on a file cabinet.

Privacy is hard to come by, and men often return to their offices at night.

Curiously, despite its being an all-male world and hidden for the most part from public scrutiny, the decorations of the carrier—those supplied, along with an occasional plastic philodendron, by the crewmen—are rather old-fashioned, decent, and modest. Where one might expect a luxuriant explosion of pornography on every wall and door, one finds instead posters of fully-clothed girl-women displaying hardly more than the shadow of a nipple pressing against a blouse or bikini and seeming to beckon one, with childish eyes and parted lips, to a romantic life of dancing and kissing, rated PG. The chaplain, for his part, distributes posters depicting sunlit landscapes and mossy brookside scenes captioned with inspirational phrases.

Perhaps nothing so much as an aircraft carrier, with its complex of thousands of interlocking lives and its self-sufficient systems of machinery, and with the evident satisfaction of most of its citizens, gives credibility to those futurist fantasies (on their face so lonely and bleak) of great cylinders sailing through space on the way to uncharted worlds, carrying populations of tens of thousands with their homes and flowers and animals and even little farms growing in synthetic sunlight, while ranges of fertile hills hang, disturbingly inverted but reassuringly distant, in the sky overhead.

THE DANCE ON THE DECK

THE PROCEDURE by which airplanes are launched and recovered is essentially simple, but it consists of so many interlocking details developed over years and from the near escapes and disasters of thousands of past landings and launches that, as one carrier officer said, "We could give the Russians this whole ship and every airplane on it, and it would be fifteen years before they could figure out how to use it." Carrier operation is one of those arcane arts, like the forging of samurai swords, that

engines and damage them. The airplanes have meanwhile been positioned on the deck so as to permit them to taxi to their assigned catapults in the proper order. After the FOD walkdown, the signal to start engines is given; soon the deck is engulfed in a deafening roar that will persist until the launch is complete. Speech is impossible; ears are heavily padded; all communication is by mouse and by hand signals.

Carriers use steam catapults about 300 feet long to launch their airplanes. Formerly,

FOD walkdown.

would be extinct except for a slender but unbroken stream of skilled practitioners who have kept them alive.

Before a series of launches begins, there is a "FOD walkdown." Several hundred deck crewmen form a dense line from one edge of the deck to the other and walk slowly from bow to stern, eyes fixed on the deck at their feet, looking for loose objects—a bit of wire, a chunk of "non-skid" broken loose from the deck surface—that could be sucked into the delicate jet

the airplane would be attached to the catapult by means of a bridle hooking on to the wings and the fuselage—and a few, like the RF-8 and the old piston-engined carrier-on-board delivery airplane (COD) still are—but most carrier airplanes are now fitted with a "nose tow" system. The nose tow is a bar protruding from the nosewheel strut, with cylindrical fittings at its front end. These engage hooks on the front of the "shuttle," the portion of the steam catapult that protrudes above the deck. A second

28

bar extends aft from the nosegear and is secured to the deck by means of an hourglass-shaped breakable link called a "holdback." The strength of the holdback is proportioned to the type of airplane.

Preparatory to a launch, the shuttle slides back along the groove in the deck. The airplane is rolled up to the end of the groove and its nosegear centered by means of a metal channel about ten feet long called the "box," which is put into place during launches and serves to center and align the

Plane captain signals start-up of his EA-6B as maintenance men and checkers stand by.

Breakaway "holdbacks" are sized and color-coded according to plane type.

nosewheels with the catapult.

The jet blast deflector rears up out of the deck. A greenshirt, the "hookup man," runs forward, hunched over, and kneels beside the nosewheel. He engages the holdback and the nose tow, which on his signal is then tensioned to a preset value of 4,000 pounds. The nosegear is now pinioned between the pull of the nose tow and the restraint of the holdback. The catapult steam pressure is set according to the weight of the airplane to be launched—a value that is

Hookup man signals "tension"; note the nose tow and the holdback.

29

verified several times by the plane crew, pilot, and cat crew.

Satisfied that he has a good hookup, the hookup man looks over to the deck edge, where the cat trigger buttons are, and sees the cat operator holding up a finger which means, "ready on the cat." The hookup man makes a whirling motion with his right hand and points forward. The yellowshirt director, standing ahead of the airplane in sight of the pilot, holding his hands in fists meaning "brakes on," now gives the signal

"brakes off, full power!"

Then the hookup man with a gesture hands the airplane off to the cat officer in charge of the pair of cats, who thrusts his right hand, two fingers extended, into the air and waves it with a rapid rotating motion. The pilot scans his instruments and moves his control stick forward and back, from right stop to left stop. Three or four troubleshooters rapidly go down the sides of the plane, checking for leaks, proper engine running, control movement, anything ab-

Left to right:
Weight board confirms pilot's weight estimate;
cat operator signs "ready on the cat";
hookup man vaults clear;
cat operator with both hands in the air, just prior to hitting the button;
cat officer returns pilot's "ready" salute;
catapult officer continues signaling for full power as he takes a last glance along the cat track.

normal. When they give their thumbs-up (the signal almost as ancient as a smile, that was already old in Rome), the cat officer, with exaggerated, almost comic movement, looks down at his hookup man, still kneeling by the nosewheel, who confirms a good hookup, springs from his crouch with a flourish as though he had just lit the fuse of a bomb, and runs clear of the plane. After looking up at the pilot, who salutes to signal his readiness, the cat officer points at the shooter at the deck edge, who has both

30

hands held in the air above his head; turns and looks forward, again with a self-conscious and premeditated motion; and then turns back to face the airplane, genuflects, and touches the deck.

As the cat officer touches the deck the cat console man at the deck edge looks aft, looks under the airplane, looks forward at the clear cat track stretching along the deck out to the sea, and then drops his hand, bends, pushes the button, reemerges momentarily to bring both hands together

while behind on the deck the crewmen, without waiting to watch the receding jet, run crouched like goblins through the swirling cloud of opaque steam.

The shuttle slides backward down the cat track, dispersing the last wisps of steam before the jet blast deflector drops back into the deck; another airplane spools up its engine and rolls forward. The hookup man is beside the nose, deftly evading the airscoop and the scissoring jaws of the nosewheel which, like those of a crocodile,

over his head, and then drops, seeming to keel over sideways from the waist, out of sight.

As he is dropping from sight, he presses a signal button telling the catapult controller a deck below to fire. The catapult almost instantly reaches full pressure, the holdback breaks in two, and the airplane, its pilot bracing his helmet against the headrest, bursts forward. It accelerates to about 150 knots in less than two seconds and 300 feet and is airborne, climbing away, banking,

would separate him effortlessly and instantly from his hand. The cycle begins again.

The launch process, which can involve as few as twelve or as many as fifty airplanes going off three or four cats, is well planned in advance; it unfolds rapidly. The airplanes wait in line, roll into place, and are prepared and shot with precisely the same movements each time: a choreography whose tempo is that of an impassioned dance, and whose unvarying sign language embodies

31

the precision, grace, beauty and equilibrium of gestures, studied and seemingly exaggerated, found in neoclassical paintings and sculptures of scenes from Roman history and the Old Testament.

Sometimes airplanes go off one of the bow cats and one of the waist cats at once, banking the moment they leave the edge of the ship, climbing away on divergent paths. As they leave the edge of the ship some make a sudden and indispensable adjustment of the elevators. The F-14s do not; they

Left: *Cat officer drops to his knee and points down the deck to signal launch.*
Center: *F-14 hurtles down* Kitty Hawk's *bow cat with afterburners lit.*
Right: *Greenshirts leap over the steaming cat to prepare for the next shot.*

set their elevators to the trim position and keep them there once airborne. But the A-7s hold full back stick—full up elevator—on the cat, and then push the stick forward as they become airborne. Sometimes a pilot

32

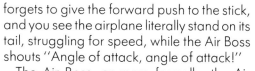

forgets to give the forward push to the stick, and you see the airplane literally stand on its tail, struggling for speed, while the Air Boss shouts "Angle of attack, angle of attack!"

The Air Boss, or more formally, the Air Officer, is the chief controller of flight operations. He occupies a cramped, crowded station on the island from which he can survey most of the flight deck (on some of the carriers he can't see the number one cat; the bridge is in his way). He and his assistant, both veteran carrier fliers, are surrounded

by radios, switches, knobs, selectors, buttons, headsets, and rows of phone receivers, all different colors and frequencies and codes; they can talk with almost anyone on the ship. During a launch or a recovery their glass-enclosed cell is full of voices, the electronic character of the line or radio hookup carrying the voice added to the timbre of the individual callers. The disembodied voices are devoid of the eerie, bored calm of a civil air traffic control center or of a telephone switchboard. A sense of

Right: Kitty Hawk's *Air Boss.* Far right: *Transparent status boards on Pri-Fly keep the Air Boss and his crew up to date on flight operations.*

vocal proportion is easily lost in the din and turmoil of the flight deck. Calls arrive in shouts and cries, sometimes difficult to understand as the caller tries to make himself heard over the thunder which, in fact, is filtered out by the transmitter, leaving only his yell surrounded by an incongruous silence. Other voices, coming from sheltered places, are calm. Sometimes angry words are exchanged; often a tone of heavy sarcasm permeates a call and is answered with measured irony. Tempers are indulged, but not usually lost; not speaking face to face, the communicators permit themselves liberties they might not take in person. Sometimes so many voices are speaking that the Air Boss can't respond to all: some he ignores; some he remembers; some he answers immediately. A continuing background of echoing voices repeats certain messages, verifying that the information has been recorded at a number of stations.

The Air Boss, his flat and drawling voice floating over the continuo like that of a solo singer, gives to his words a certain dramatic emphasis, employs certain phrases, and indulges in certain locutions that are the badge of his preeminence, his star status: only he is allowed to waste words. His assistant engages him in a continuous dialogue:

Going amber in six seconds. (Amber is the ready light signal on the island, signalling the imminence of launching. Great emphasis is placed on starting each exercise at precisely the appointed second.)

Four oh five is changed to four eleven, Boss. (The Air Boss is addressed by everybody as "Boss," which gives Primary Air Control the sound of an old plantation.)

Four oh five is now Four eleven.

That would be that damn cat down, we've got five goddam fighters. (The Air Boss: only two of the four cats are equipped with an oversized jet blast deflector and can launch F-14s; one of these isn't working.)

34

Amber!

Can't you retract that thing on the retract engine?

We've got the tractor standing by in case it's stuck.

The shuttle on the broken cat won't retract; a tug is brought and drags it back to the start position.

Okay, we've got forty-five seconds to red. Winds are good. Thirty seconds.

CD-1, approach, CD-1, CD-1.

Ten seconds. Red. Green on the fifteen-second mark. Green!

Stand by! Stand by! Stand by!

There is a roar and a clatter as the first jet, an A-7, is shot from a waist cat, a slender filament of steam preceding the shuttle down the track and being swallowed up, as it goes, by the gulping mouth of the jet. The Air Boss looks at his wristwatch with a satisfied air.

Six three two airborne.

That's getting 'em off on the stroke. On the second.

I'm not sure the guy on cat four appreciated that.

I don't give a damn if he did. If he can't stand the heat, he better get the hell out of the kitchen. It ain't no game for boys.

Stand by! (Thunder of engine.) *Cat two!*

A loudspeaker intones: *All F-14s will be going military power off cat two.*

Boss, five oh six has been cancelled.

Cat four, four eleven, stand by! (Roar, thud: a launch.)

Four eleven airborne!

The process is automatic; the airplanes follow one another into the air at intervals, like bullets from some slow-firing gun.

One sixteen airborne!

Three zero zero airborne!

The way I count it, five, six to go.

What's the problem with the fighters down there? Is two oh one down?

Six to go, boss?

Two oh one's down, in that case, one two, three . . . five to go, I guess.

Stand by, three oh four!

Three oh four airborne. (Voices repeat like an echo: *Three oh four, three oh four . . . airborne . . . airborne . . . .*)

One by one, repeating exactly the same ritual, the jets roll onto the cat, the preliminary dance unfolds, and they are shot into

flight. Then as the second-to-last launch is underway, a voice brings in a new phrase:

Six twelve, ten miles.

A returning jet, its weaponry spent and excess fuel jettisoned, is ten miles astern, lined up, travelling three miles per minute. The launch continues:

I take it one two one's down.

Two oh one's not down.

Two oh one's down. Two to shoot. Bridge, Primary, two to shoot.

Stand by.

Waist, I'll need a fast cleanup, we've got an aircraft a mile and a half, correction three miles now.

35

Yessir. Okay, you've got 'em? Going to sixteen. (He is switching to radio channel 16 for the recoveries.)

Stand by!

Personnel on one fourteen, get off the aircraft, we're starting a recovery! Get off the aircraft, one fourteen! (The Air Boss is shouting peevishly on the deck PA system to two enlisted men lounging atop the grey haunch of a parked F-14. They look up dully and with extreme reluctance begin to move.)

Seven oh six airborne!

How far's the first aircraft? (The Boss is speaking.)

Overhead. (This cannot be the answer.)

How far's the first aircraft?

I'm trying to get the Boss. (Still wrong.)

How far's the first aircraft?

Four miles, Boss.

Four miles? Okay.

Stand by, cat two, two zero zero!

Cat two, two zero zero.

Okay, they've got to hot-pump, crew switch six oh four at this time.

I need six oh four if I'm gonna hot-pump, crew switch it. Where is he in the stack?

Okay, launch complete! Ready forward. Between decks, Primary.

Can I wrap up the bow?

Launch complete.

No, no! We've got a hot-pump.

That's right, we've got to hot-pump 604. Hold the bow open. (Hot pump, crew switch: a Hawkeye, a turboprop patrol and radar picket plane, is coming in; it will land, be refuelled, get a new crew, and be launched, all with the engines running. In principle, it should be the first plane in, so that it can be launched as soon as possible; but it isn't. The Boss is unhappy. In the meantime, against the grey sky astern, a darker grey point has materialized, grown larger,

Carrier Air Traffic
Control Center
(CATCC).

become an RF-8 Crusader, swelled rapidly, and then roared away without ever touching down.)

Wave off, foul'd deck! (He has been waved off because the deck, still being cleared after the cat launches, is not ready to receive him.)

Soon as we get him on cat two we'll pump him and man him and shoot him.

No problem with that.

Foul'd deck!

He's at fifteen miles inbound.

What's inside of him?

The other four aircraft. I got word that we had to turn him around as the first aircraft was approaching the ball. I can't reshuffle them, I can't perform magic. (This is a calm voice over the radio. Another radio voice from a different part of the ship offers with defensive sarcasm:)

Neither can we.

The Boss is getting angry; things will not go so smoothly this time. His voice is a decibel or two louder than the rest, and more final: *And neither can we, bud, so how about getting with the damn work, awright?*

When a launch is complete, the deck is done with it. The airplanes are connected to the carrier through CATCC—the Carrier Air Traffic Control Center, located in several dark and heavily curtained rooms deep in the core of the ship, where area radars display the creeping dots that represent airplanes on maneuvers. Transparent status boards glow with colored writing, the luminous letters taking shape, like rapidly moving worms, under the crayons of men trained to write in mirror-image on the back of the boards. Lists of aircraft are compiled, updated, juggled, erased, and replaced. Voices, live and on radio, call out the order

of returning planes. In an adjoining room, other boards and charts display strategic information, and on a pedestal are two lordly chairs from which captains and admirals would direct the war games or the wars, and in which sprawl, more commonly, drowsy petty officers with rumpled jackets and dull shoes.

Primary and the cat officers control launches, but recoveries hinge upon the direction of the LSO, the Landing Signal Officer. Descended from the man who stood at the stern of the old carriers and with flag signals directed the approaches, the modern LSO is a critic. He is a veteran pilot, and a good one; and he has seen thousands of landings from the little platform surrounded by canvas nets on the aft port side of the ship. He grades each approach and landing; after each recovery he critiques the returned pilots. Usually they don't argue with him: it's of no use. It may seem strange that he can judge the approach better than they can, since it is they who see the "ball"—the optical glideslope—the lights, the glidepath, and angle of attack indicators; but he can, and he is the final authority. Red, yellow, and green lights on the nosegear struts of the airplanes give him angle of attack information; he listens to the rise and fall of the engine whine; he guesses what is going to happen before it happens. In his hand he holds the "pickle switch"; if he doesn't like the way an approach looks, he presses a button and a cluster of red lights around the ball waves the pilot off. He is the man who, in the swift and terrifying final moments of night recoveries, keeps the pilots "off the ramp"—keeps them from dropping too low and striking the aft edge of the deck.

The accuracy of the LSO's eye is remarkable; many are said to be able to detect an error of five feet in glidepath height at a distance of half a mile. Lineup is more difficult to judge; and at night, when the real horizon is invisible and the airplane is only a cluster of moving lights, *everything* is more difficult to judge. Nevertheless, it is the LSO, depending on information from the airplane's lights, verbal reports from the CATCC, and most of all upon his eye and trained intuition, who is the final judge of the quality of an approach. It is he who gives the pilots their marks, which he keeps in a confidential log; and in the long run it is his decisions that make and break navy pilots. Marks run from an underlined (but still rather noncommittal) "OK," meaning a perfect pass, to an equally noncommittal "C," meaning an unsafe approach with gross deviations.

Navy pilots are egotists; instructing them day after day requires tact. Even the conversation between the LSO and the pilot during the approach involves a delicate touch. There are good LSOs and bad LSOs. A good LSO looks out at the incoming airplane, sees a trend, listens, hears power coming off, and says "Don't go low," or "Power." He anticipates a problem and offers a corrective that the pilot at the same moment may be offering himself. A bad LSO in the same situation says, "You're low." The pilot feels insulted; he *knows* he's low. The LSO then becomes an adversary. He should, ideally, be an alter ego, a part of the pilot himself, so that the LSO speaks to the pilot in the same tone, at the same moment, as the pilot speaks to himself. The psychological tone must be right, in order for the pilot to internalize the observations of the LSO and learn from them. If the pilot feels resentful of the LSO or resistant to his words, he won't learn.

There are usually at least half a dozen people on the LSO platform: the duty LSO, perhaps one or two other LSOs who are learning the trade, fellow squadron members, and one or two radio listeners and talkers who make calls for the LSO, relaying them from various sources on the ship. Everyone is aware of the distance separating him from the nets that surround the platform; that is where they will all go, helter-skelter, if a plane aims at them rather than at the centerline, or appears about to strike the ramp. The arresting cables are often hidden from their view by airplanes parked on the port aft corner of the deck (which in most carriers is the port elevator). As a plane swoops down to the deck in the last second of the approach, everyone on the LSO platform spins around and crouches to see, underneath the parked planes, whether the approach ends in a "bolter" — a miss — or a "trap." Conversation is a mixture of shouts and murmurs, modulated by the surrounding noises and the growing suspense as a plane approaches. The LSO has a telephone on an open line shared by the Air Boss, the CCA (Carrier Controlled Approach) controller in CATCC, and the pilot; in the other hand he holds the pickle switch.

It is night.

One mile.

The LSO's talker calls out: *Number one, three oh four, all down, clear deck. Clear deck!* (The first airplane on approach is 304; his gear, flaps and hook are reported down by CATCC, and the deck is clear to receive him.)

Over the radio to the Boss, the LSO, and

39

the pilot comes the voice of the CCA controller, who is watching the approach in profile and plan view on his radar scope. *Three oh four on line very slightly right, three-quarter mile, call the ball.*

The pilot replies, when he has the ball in sight: *Three oh four, center ball, four point eight.* (He has the ball, can see the carrier's lights, and has 4,800 pounds of fuel remaining.)

Lights in the darkness: the LSO sees him coming, hears the engine, forms an airplane in the black wake of the ship, thinks it down, only to see it roar back into the night after touching the deck.

Attitude. (Explains the LSO.) *He just has to loosen up a little bit more.*

I think you've got one closer there. What have we got here, no wing lights?

Number one, sir. Check wing lights —three one zero, check external lights on please.

Suddenly the red and green wingtip lights appear.

Okay.

Three one zero, all down, foul deck. Clear deck. Clear deck!

(The CCA controller:) *Centerline right three, going above glidepath, one mile, three one zero, left two, coming down, slightly above glidepath, three one zero slightly above glidepath, on centerline, call the ball.*

Three one zero on the ball.

Roger ball.

The jet materializes out of the blackness, swells, thunders on the deck, and bolts.

Oh hell!

Laughter, shouts. On the platform tension

The LSO platform: top to bottom: *hook check; at the ramp; trap.*

Aircraft's last-second corrections are visible in this time exposure of a night trap aboard Ranger (CV-61).

breeds quietly. The CCA controller again: *Three one zero, pick it up on heading two niner zero, climb to angels one point two.* (He is to head 290°, climb to 1,200 feet, and reenter downwind.)

Number one, two zero seven.

An F-14! He'll show you how to get aboard.

The half dozen people on the LSO platform, observers, assistants, and apprentices, all talk at once.

Where's that wire, we'll find it!

Did that guy trap or not?

He got all four.

They're really messin' up.

Hey, they looked good last period.

Two oh seven on glidepath, on centerline, one mile. Call the ball.

Two oh seven, on the ball, eight zero.

Roger ball, eight oh.

With the undulating whine of approaching engines, the huge, angular F-14 appears in the faint light, elevators flapping, attitude light yellow, hits the deck at full power, and traps.

From the standpoint of the ship's crew, as well as that of the pilots, the recovery of aircraft is the climax of carrier operations; it is the eye of the needle through which they must pass. By the general consent of those involved and those who only observe, it is the most difficult trick in aviation.

The angled deck, on which landings take place, is about 500 to 700 feet long, depending on the carrier. Its approach end,

41

the aft end of the ship, is called the ramp or rounddown; about a third of the length of the angled deck forward from the rounddown is the hook aim point for landing.

The hook aim point is at the center of a set of four arresting cables strung across the deck at intervals of about forty feet. The central segment of each cable, called the cross-deck pendant, takes the shock of the arresting hook engagement. About one and a half inches in diameter, it has a tensile strength of 176,000 pounds. Regular inspections for frays are made during its short life, and after every hundred traps it is removed and thrown over the side so there is no chance of a used wire being reinstalled.

An F-14 of Fighter Squadron 111 heads for the two-wire on Kitty Hawk.

The cross-deck pendant is secured at either end to cables that pass around drums and down into the space below the deck, where they run through reeves of pulleys to hydraulic dampers and to the retraction motors — an amazingly intricate mechanism that will stop an aircraft in a precise number of feet, regardless of weight and speed.

About two-thirds of the way down the left side of the deck on a cantilevered platform is the "meatball," or Light Landing Device. This contraption, a descendant of more primitive optical glideslopes, is gyro-stabilized so that, within sea-state limits, the glideslope remains stationary in space while the ship pitches and rolls. It consists of a set of five Fresnel lenses in a vertical array surrounded by reference lights, and ground in such a way that if the pilot is on glidepath he will see the middle lens illuminated; if he is high, he will see one of the top lenses; and if low, he will see one of the bottom lenses. When the pilot "calls the ball" in the last mile of an instrument approach, he is verifying that he has visual contact with the ball. If he doesn't see the ball, he simply says "Clara."

The meatball is set to provide a glide-slope angle of about three degrees, which is variable within narrow limits depending on sea-state. However, it is the pilot's eyes that fly down the optical glideslope; in the on-speed approach attitude, the hook — the part of the airplane to be placed between the two and three wires — is as much as twenty feet below the pilot's eyes, the exact "hook to eye value" varying from type to type.

The Light Landing Device or "meatball."

42

Since the goal is to have every plane put its hook over the ramp at approximately the same height—at least ten feet, and preferably sixteen on the big-deck ships—it is necessary to alter the height of the glideslope over the ramp. To change the glide angle would be undesirable (airplanes arriving at too great an angle would hit too hard, ones arriving at too shallow an angle would be in danger of hitting the ramp). One solution would be to move the LLD fore and aft, or vertically, along the side of the ship; but this is structurally impractical. Instead, the entire LLD rolls on an axis parallel to the runway centerline. Since it is located at the edge of the angled deck, this has the effect of raising or lowering the portion of the fan-shaped glidepath which lies over the centerline. Thus, the proper approach angle is maintained, but the height at which the glidepath crosses the ramp is changed.

The LLD is mechanically operated under control from the LSO platform; if it goes out of service, a backup system similar in operation but manually controlled, the MOVLAS, takes over. The MOVLAS is situated on the deck edge a little forward of the LLD.

The LLD must be adjusted to compensate for a number of variables, including ship's trim in pitch and list. Small angular variations produce surprisingly large differences in touchdown point. For example, one degree of pitch on the deck moves the hook touchdown point over 120 feet—three wires. These adjustments are programmed into the LLD by the LSO. Touchdown point and hook-to-ramp clearance can also be affected by the angle of attack of the approaching airplane; a long airplane like the F-14 can pitch nose-up and catch a one-wire—the first wire—even though the pilot

is seeing a high meatball. The tolerances in every respect are incredibly small.

The ball provides the approaching airplanes with vertical guidance; lateral alignment is provided in daytime by the perspective of the deck, and at night by the lines of the lights down the centerline and on either side of the runway and by a "dropline" of orange lights which hangs down the stern, on the centerline. The purpose of the dropline is to help provide a sense of perspective; one of the problems of approaching the carrier visually at night is that the little cluster of lights in a vast space of black velvet does not readily translate itself into a horizontal carrier deck pointing in a certain direction. The eyes and the brain, supplied with minimal cues for forming an imaginary picture of the ship, may instead see a hole surrounded by vertical lights, or simply an undecipherable pattern. The tail of the dropline, which the pilot knows to be vertical and central, helps him to interpret the vague clues of the deck lights: if the dropline is aligned with the centerline, he knows he is on center.

In addition to visual cues, the carrier provides approaching aircraft with two kinds of electronic guidance. The first, ILS (Instrument Landing System), is similar to civilian ILS, and beams an electronic glideslope behind the ship which is received by the airplane's radios and converted into a crosshair display on the panel. The second is a radar system coupled with a computer which locates the approaching airplane, computes its position with respect to the desired glidepath, and converts this information into the same crosshair display for the pilot as is used for the ILS. On a "coupled" approach, the computer can also fly

Hornet heads for a solid three-wire trap.

the airplane, through the autopilot, to a fully automatic landing —without the pilot touching the controls.

Finally, the ship has TACAN. This radar system converts the travel time of a coded pulse between ship and airplane to measure the distance separating them, and also gives the pilot an exact bearing to the TACAN station.

Three types of approach are used. In Case One, with a 5,000-foot ceiling or bet-

ter, and visibility greater than three miles, the airplanes locate the ship visually. They fly an upwind leg in formation abeam the ship, and peel off one by one to execute the classic 360-degree overhead approach, joining the glidepath only a third of a mile behind the ship from a low-altitude, flat turn. The approach is visual from the outset; the only radio navigation involved is that required for locating the ship in the first place.

A Case Two approach is used in cloudy weather when ceiling is 1,000 to 5,000 feet and visibility is three miles or more. The arriving airplanes circle or "marshall" twenty to thirty miles behind the ship with visual separation, below the clouds. Usually, different types are marshalled at different altitudes. They approach visually, one at a time, or, in some cases, in small formations of two to four aircraft.

In Case Three, the weather can be as bad as 200-foot ceiling and half a mile visibility, or even, if no alternative landing fields are available, zero-zero—no ceiling, no visibility. The airplanes marshall individually and hold behind the ship at stacked altitudes, the holding TACAN distance usually being the altitude in thousands of feet plus fifteen: 5,000 feet and 20 miles; 6,000 and 21 miles; 7,000 and 22 miles; and so on. They fly an oval pattern, maintaining 1,000-foot intervals, with orders to depart the holding fix at a certain time. If a pilot is to leave his fix at thirteen minutes after the hour, he is to do so at exactly that moment, as the second hand passes the twelve. Approaches normally begin at one-minute intervals.

The pilots plan their flight pattern for an on-time departure from the marshall fix by modifying the shape of their holding patterns. Occasionally, if they have gotten orders to marshall unexpectedly early, the traffic arriving at marshall may be chaotic: jets careering in at 400 knots and then dropping brakes, gear, and flaps to slow down at the fix, while others are circling tightly, or adding bulges and lobes to their holding patterns to kill an extra twenty seconds. If they are above the clouds they can see one another; the holding patterns are larger than the one-mile graduations from the ship, and each pilot sees, a thousand feet below him, the airplane that will precede him to the ship, and, above him, the one that will follow him.

In theory the holding pattern is oval, with half of a four-minute turn at either end and straight legs a minute or so long in between. A four-minute turn is one that takes four minutes to cover the full circle; thus each end of the pattern uses up two minutes. If a pilot arrives at the holding fix—represented by a certain radial and TACAN distance from the ship—seven and a half minutes before his expected approach time, figuring four minutes for the turns, he has to kill three and a half minutes on the straight legs, or one minute forty-five seconds per leg.

There is, however, a complicating element: the ship is moving away at an unknown speed, compounded of its speed in the water and the speed of the wind not on the surface (where it will be about twenty-five to thirty knots with respect to the ship) but at the altitude, say 7,000 feet, where the plane is holding. The pilot has to make a guess at this speed, unless he holds long enough to time several legs and deduce it empirically. If, say, the wind and the ship's motion together add up to thirty knots, then the ship will be one mile farther along every two minutes. If the pilot has to hold for eight minutes, the ship will have moved four miles in the time since he began his hold. The pilot is travelling at 250 knots and covers four miles in one minute. So he must shorten his straight legs by thirty seconds in order to be at the TACAN fix at the right instant.

Naturally, some leeway is available. The pilot can begin the approach on the second whether or not he is at the exact nominal TACAN distance; if he starts a little farther out than he intended, he can make up the difference during the dirty-up-and-slow-down portion of the approach, carrying a little more speed a little longer.

45

When the pilot leaves the final fix, twenty miles or more behind the ship, he is flying 250 knots—nearly 290 miles per hour. He covers the first ten miles in a bit over two minutes, descending at about 2,000 feet a minute. At ten miles, he "dirties up": drops his gear, flaps and hook. At five miles he is in the landing configuration, slowed to his approach speed—120 to 140 knots for the jets, depending on type—and level at 1,200 feet. He is reading his speed in terms of angle of attack—the angle of the wing to the flight path of the airplane, which determines the wing's lift and is independent of the airplane's weight. His ILS needles are his primary position reference, along with his TACAN distance; at this point the vertical needle, representing lineup, should be centered, and the horizontal one, representing glideslope, should be at the top of the instrument, because he is flying level underneath the glideslope, waiting to intercept it. In his headset, the voice of the shipboard radar controller gives him distance, relation to the glideslope after he intercepts it, and small heading corrections: "Two miles, right two, slightly above glidepath."

In principle, the minimums for Case Three are 200-foot ceiling and half-mile visibility. In the nature of weather and of the motion of the ship under a ragged cloud deck, these figures are approximate, but they would mean that the pilot would catch sight of the ship about half a mile, or fifteen seconds, behind it. Below these minimums, in peacetime maneuvers, operations are suspended. In "blue water" operations, where there is no onshore airfield available for airplanes unable to make the carrier, airplanes would not be launched in marginal weather. In wartime, it might be another matter; aircraft could be recovered under any conditions.

In the worst situation, there is always the ACLS—the Automatic Carrier Landing System. ACLS combines the readings of the precision radar with the airplane's autopilot and flies the plane automatically, hands off, right down to the deck. Pilots have mixed feelings about the ACLS. Some, especially those who have flown F-4 Phantoms on automatic approaches, are confident in the system. A-7 pilots, on the other hand, are suspicious of it. The reason is partly the difference in flying qualities between the two airplanes. The F-4 is an extremely stable airplane, and changes of power setting do nothing to disturb its pitch attitude; the A-7 is much less stable, and furthermore its pitch attitude is affected by power—disconcertingly so, since the power of its fanjet engine arrives a short time *after* the pilot has moved the throttle. Perhaps the best ACLS performer is the F/A-18; with its digital avionics and computer flight controls, it follows the ship's electronics with cybernetic near-perfection. In some airplanes there has been reason to distrust the ACLS: improper placement of the airplane's radar reflector has permitted the radar to lock on to a different part of the airplane, producing a sudden jolt of the controls; and sometimes the radar has lost lock on the airplane and picked up falling rain instead, producing a violent "*Pull up!*" signal to the autopilot.

The aim point on the deck is the center of the cluster of four arresting cables; the trap of choice is the three wire. On a big-deck

F-18 Hornets of reserve squadron VFA-303 out of NAS Lemoore, California.

46

carrier—those in the 80,000-ton, 1,000-foot-long class—a two wire can be an acceptable pass, if the pilot comes in with good attitude, not pumping the nose, and holding the glidepath and speed with good power control. On the older, smaller carriers, even a one wire was acceptable, although the first wire was only 180 feet from the rounddown. The nearer the rounddown one hits, the more dangerous the approach—the ultimate danger being that of hitting the ramp, breaking apart, bursting into flames, and scattering the deck with fragments of burning aircraft and weaponry. A four wire is not good—it means the plane was high—but at least it is safe. A jet takes only about half a second to cover the distance from the first wire to the last.

As the pilot is about to hit the deck, he pushes the throttle fully forward. An instant later, if he traps, he will bring it back to idle thrust; if he bolters or "bolts," he will already be off and away, to proceed ahead of the ship, make a left-hand turn, and reenter the landing pattern for another attempt.

Trapped, the airplane is dragged to a halt in 300 or 400 feet, sometimes coming to

within a plane's length of the front edge of the angled deck and the sea. Power at idle, feet off the brakes, the pilot waits; the plane rolls back a few feet because of the elastic rebound of the cable, which, though it is an inch and a half of stranded steel, coils up like a twisted rubber band. The cable falls from the hook, the director signals, and the jet taxies clear of the runway as the next airplane is half a minute from touchdown.

They march aboard this way, one after another, each airplane rapidly approaching the stern as the one before it taxies into the crowded deck space ahead of the island that serves as a temporary collecting area for all the airplanes. Once the recovery is finished, they will be towed aft one by one.

Trapped F-14 is signaled off the deck.

It is because of this temporary jamming of the forward deck that launches precede recoveries; after a recovery, the bow catapults are temporarily useless. When the next launch closely follows a recovery, it normally goes off the waist cats. This occasionally happens in the case of oddball missions such as a recon bird going out to photograph a Russian trawler or an E-2 launching to position himself for the next exercise.

Sometimes the tailhook fails to drop the cable, or the hook won't raise, or it will raise and then not lower if the cable remains engaged. Then men run out with big steel bars to pry the cable free and guide it along the deck. Undue delay results in a fouled-deck wave-off: as one airplane, finally released, struggles to turn its nosewheel away from the ocean and toward the parking area, the next airplane roars indignantly past overhead. A wave-off is not trivial; it wastes time and fuel, and the frustrated pilot, having already dumped fuel for the approach, may, if he bolts on his next attempt, have to refuel in flight before trying again —wasting still more time and fuel. And so the deck crew works hard to keep the airplanes coming aboard. It is a matter of fuel, time, and the ire of officers; it could quite possibly be a matter of life and death.

Wave-off—fouled deck.

49

ARMS . . .

AIRPLANES FORM carriers, and in turn, carriers form airplanes. The airplanes that have operated from carriers have often been extraordinary ones, expressing the conflicting requirements of speed and weight carrying, and the controllability, low-speed flying qualities, and strength necessary for carrier operations. World War Two produced the Hellcat, Corsair, and Bearcat fighters, as well as the Helldiver and Avenger. The postwar period brought into being a classic piston-engined light attack airplane, the Douglas AD-1 Skyraider, which was still in service in the Vietnam War. The postwar period also brought new requirements: carrier-based medium bombers capable of carrying nuclear weapons, sophisticated sub hunters, and jet fighters. There was a trend toward heavier, larger, and more complex airplanes—there always is—but one of the most successful postwar designs, and one of the most durable, was the little delta-winged Douglas A-4, which was so small that its wings didn't even need to fold for carrier storage. At the other extreme were the A-3 and A-5—huge, heavy bombers—the latter called the "Vigilante."

During and after the Korean War jet airplanes with slow-responding engines and high approach speeds operated from straight-deck carriers; accidents were frequent, and there was continuous pressure for larger and larger carriers. It was this period that brought into being the supercarrier with its thousand-foot length and an-

In the Korean War, jets operated precariously from World War II-vintage straight-deck carriers like the Bonhomme Richard.

52

gled flight deck, as well as a number of novel airplanes such as the Vought F-7U Cutlass, a tailless design, and the tailless delta-winged Douglas F-4D Skyray. The most successful navy fighter of the Korean period, however, was the more conventional Grumman F-9F Panther—the forerunner of the F9F-8 Cougar, which served as a trainer for more than twenty years.

The Chance Vought F-8 Crusader, whose design was begun in the early fifties and which exceeded Mach 1 on its first flight, and the McDonnell F-4 Phantom were the fighters that saw the navy through the latter half of the fifties and the Vietnam War. They were closely followed by the generation of airplanes now in service: the F-14, A-6, A-7, F/A-18, E-2 and S-3, as well as other models in specialized roles.

The F-14 Tomcat, the fleet's primary air defense interceptor, is a variable-geometry airplane, a type pioneered by the trouble-plagued and controversial General Dynamics F-111: its wing sweep can be varied in flight from an almost straight position for takeoff and landing to a maximum sweep of 68 degrees for supersonic dashes. An over-sweep position of 75 degrees is provided for deck storage. A long, slender, straight wing is the most efficient producer of lift at low speed and also permits the use of high-lift devices, such as trailing edge flaps and

Top: *F-14 Tomcat, the navy's current fleet defense fighter. Its powerful Phoenix missile system can intercept multiple aerial targets at distances of a hundred miles and more.* Bottom: *Tomcat escorts Soviet "Bear" recon bomber over the Mediterranean.*

53

leading edge slats, to increase the load-carrying ability of the wing.

Roll control is provided by spoilers on the wing outer panels and by differential motion of the slab tailplanes. As the wings are swept, two narrow-delta canard surfaces called "glove vanes" automatically emerge from the highly swept leading edge of the wing center section, or glove, above the engine air inlets; these compensate for aft center-of-pressure shift and relieve the tailplanes of much of their balancing function, leaving them free to provide roll control.

Although it can be manually overridden, wing sweep is ordinarily programmed automatically by the onboard computer, as is deflection of maneuvering flaps above certain angles of attack. Landing flap, on the other hand, is manually selected by the pilot. Two vertical surfaces are used, as is increasingly common on twin-engine supersonic aircraft. Each vertical fin tilts slightly outward from the centerline of the engine nacelle; the two widely spaced nacelles are joined by a flat bridging structure that could be thought of as the back end of the fuselage or as part of the wing centersection. At the trailing edge of this flat body are the speed brakes, which deflect upward and downward but are not very large compared to the rest of the airplane.

The Navy's F-14As use two Pratt and Whitney TF-30 afterburning turbofans of about 20,000 pounds thrust each. These engines have been inadequate and trouble-prone from the start, and fleet pilots have long looked forward to the improved F-14D due late in the decade. It will carry superbly reliable General Electric F-110 engines, as well as modern digital avionics and count-

less improvements in the radar, weapons, and countermeasures systems.

The Tomcat's engine exhaust nozzles, resembling elaborately petalled flowers, expand and contract to produce optimum pressure ratios. Huge, canted two-dimensional air intakes flank the fuselage, and the landing gear, which is adapted from that of the A-6 Intruder, retracts into the nacelle and wing glove just outboard of the engine. Over 16,000 pounds of fuel—2,500 gallons—is carried in internal tanks.

The two crew members, pilot and Radar Intercept Officer (RIO), sit in tandem seats under a long bubble canopy. Both have Martin-Baker rocket-propelled "zero-zero" ejection seats (meaning they can be successfully ejected at a standstill on the ground). The RIO has no flight controls, though he

54

F-14 crew awaits launch on Kitty Hawk.

uses a joystick similar to the pilot's to aim the radar. Looking at two cathode ray tubes that can be used for either navigational or radar displays or to pick up information from a ground data link, the RIO can watch, for instance, the progress of an approach on a repetition of the ship's Precision Approach Radar. Using the target acquisition and tracking mode, he can vector the pilot to the target, or he can take over control of the airplane through the autopilot and fly it electronically according to computer commands. But since he cannot see directly forward outside the cabin, he takes the critical final moments of the carrier landing on faith.

Although the F-14 depends heavily on its computer system in all aspects of its mission, its control system is not purely electronic, or

"fly-by-wire"; instead, hydraulically boosted controls are used in the tail surfaces, and a fly-by-wire system for the spoilers.

The maximum takeoff weight of the F-14 on land is 72,000 pounds; more typically, it operates from carriers at about 58,000 pounds. Its approach speed is 120 knots plus one knot for every thousand pounds of fuel aboard. Maximum speed is given as Mach 2.34, minimum as less than 75 knots.

Contrary to the imaginings of civilian pilots who have always longed to take the controls of a fighter and who have sighed vicariously over the silky ailerons of a Spitfire or the sheer power of a Bearcat, many of the current fighters do not have particularly delightful handling qualities. It isn't that they don't fly well. Pilots are unanimous in their praise of the F-14's uncanny low-speed maneuverability, its tolerance of high angles of attack, and its remarkable ability to climb almost vertically after takeoff and then nose over sharply at minimum speed and fly away level without loss of altitude. The F-14 can be controlled at an angle of attack of almost 80 degrees—unheard of in any earlier fighter.

But there is something mysteriously missing in the control systems of this generation of fighters, with their hydraulic boosts and fly-by-wire controls: they give no feel of the airplane. Instead they have "artificial feel." It would seem simple enough to program ideal stick forces into a largely electronic airplane, but the F-14 uses bungee springs to provide feel and there is a dead area in the middle where nothing much happens

55

when one moves the controls. Beyond the dead area is the realm of control authority; there is an abrupt rise in resistance—the controls are in fact extremely heavy—together with a nonlinear increase in roll response: first nothing, then too much. All this one gets used to; but the sense of pilot unconsciously blending with aircraft, of the complete internalization of the act of flying, is harder to achieve. It used to be said, as the highest praise of an airplane's handling qualities, that you had only to think about rolling it and you were already halfway around. That common remark was not so much a humorous exaggeration as a precise description of the sense of identity with the airplane that a pilot can have when the control forces are light and yet precisely proportional to the reactions.

The F-14 is, in any case, a spectacular performer despite its less-than-satisfactory engines. It can take off in a quarter of a mile, accelerate to 400 knots in less than a minute while climbing steeply, fly at almost two and a half times the speed of sound, and despite its great size and weight, can dogfight and outperform much smaller and lighter fighters. None of this matters, however, on the carrier approach. There what counts most is stability, freedom from pitch responses to power or Direct Lift Control (DLC) spoilers, a good sink rate, a low approach speed, and the best possible coordination between the pilot's mind and the airplane. In all these respects, the F-14 is no better than average. Like most other carrier aircraft, the F-14 corrects its aerodynamic shortcomings with novel, elegant, and amusing systems, such as artificial stability augmentation, autothrottle, DLC, and multiple levels of roll authority.

The F-14 is a floater; even dirty it's too clean because of its huge wingspan of over sixty-four feet; if it goes too high above the glidepath it sometimes can't get down without picking up speed even with the engines at idle. Thus the need for the DLC, an intermediate or "floating" position of the spoilers. A spring-loaded thumbwheel on the stick controls them; together they can be lowered to increase lift, or raised to dump it, without any change in power setting or pitch angle.

The neutral roll stability and nonlinear roll response of the F-14 require variable roll authority; in fact, as on old taildraggers, the rudders rather than the spoilers are the best roll control during the last stage of the approach. Pilots make one novel use of the roll spoilers: by waggling the stick rapidly from side to side at the last moment of the approach, they can make a small downward adjustment in the flight path since the spoilers work, as their name implies, by destroying lift, and the airplane suffers a small loss of altitude everytime a spoiler is raised.

The F-14's speed-holding ability is poor; hence the APC or autothrottle, a speed-holding autopilot that senses changes in airspeed and moves the throttles accordingly. Rather than have to fly throttles for altitude and pitch angle for speed, the pilot simply uses the stick—his pitch angle control—to chase the glidepath, and the throttle and airspeed take care of themselves. One would think that pilots would become dependent on a crutch like APC and lose their ability to land well without it. But the navy says that when F-14 pilots have to make an approach without the APC they do as well as with it; and some leave it off as a matter of course.

Less sophisticated aerodynamically, but highly advanced in its electronic systems, is the A-6, also a Grumman design, which entered service with the fleet in 1963 and was responsible for much of the damage inflicted upon Vietnam. The A-6 exists in a number of versions and roles. The original concept was an all-weather subsonic tactical bomber that could deliver up to 18,000 pounds of bombs, conventional or nuclear, against targets completely obscured by weather and darkness. By comparison, the B-29, the largest bomber used by the U.S. in World War Two, carried 20,000 pounds of bombs; but it had a wingspan of over 140 feet to the A-6's 53, and an empty weight of over 70,000 pounds to the A-6's maximum

A-6 Intruder is the navy's all-weather carrier attack bomber. This sophisticated brute can carry nine tons of ordnance and can deliver them in conditions of zero visibility.

takeoff weight of 60,000 pounds. The A-6 lacks only the B-29's range, but makes it up in being able to operate from carrier decks and refuel in the air.

This remarkable jump in weight-carrying ability was brought about principally by the introduction of the jet engine. Jet engines have a phenomenal ability to consume air and fuel, and so to produce power. Thrust and horsepower are not directly comparable, but at about 300 knots a pound of thrust is equal to one horsepower; so the F-14 climbing at 300 knots in afterburner is developing about 40,000 horsepower. Seven F-14s are as powerful as the carrier itself. By comparison, the largest piston engines used in World War Two could produce 3,500 horsepower and then only for short periods.

Two other innovations have also been significant: the introduction of guided or "smart" bombs which almost unerringly

57

find a target, making unnecessary the shot-gun approach to bombing used in the for-ties; and the shift from internally to externally carried stores. Bombs, missiles, and auxil-iary fuel tanks are now carried outside the fuselage, usually on racks beneath the wings, with several effects: flight stresses on the wing are less than when all loads are carried within the fuselage, ordnance loads can be more quickly disposed of in an emergency, and a greater variety of ordnance can be carried and used in any sequence. Thus, while they carry the same destructive force as the largest World War Two strategic bombers (even omitting nu-clear weapons from consideration), today's attack aircraft are much smaller; and though they suffer a considerable drag penalty due to their external stores, once they have unloaded those stores they are far smaller, lighter, and swifter than their predecessors.

The abilities of the advanced versions of the A-6, particularly the A-6E Intruder and EA-6B Prowler (the Electronic Countermea-sures or ECM version), are awesome. Their electronic equipment is extraordinary. Radars, infrared sensors, laser ranging de-vices, and inertial platforms supply informa-tion to computers that identify and track targets on the ground or in flight, or create a synthetic terrain picture on the eight-inch Vertical Display Indicator. The VDI is a TV screen that not only displays the unseen terrain, permitting the pilot to fly at low level among mountains as if he could see them, but also generates a command flight path ahead of the airplane for the pilot to follow.

EA-6B Prowler, expensive and compli-cated four-seat version of the A-6. Its wings and tail bulge with pods and anten-nas that detect, confuse, and neutralize enemy electronics.

58

All these synthesized terrain features move as they would if the weather were clear and the pilot were looking at them through the windscreen; the computer updates the display ten times per second. The pilot manages his weapons systems entirely through the computer, selecting, arming, aiming, and firing ordnance in any combination.

The A-6 is an ugly airplane, and its popular names—Prowler, Intruder—have ugly connotations. The namers of the airplane seem to have been vaguely uneasy at the powers of the monster—at the thought of how, out of the low hanging clouds that in any other war would have brought a respite of bombing, out of the rattle of wind and rain on the palm leaves, there could come

the scream of jets, low but invisible, and tons of fire and bombs falling out of nowhere with unerring accuracy.

More conventional, more comfortable to minds whose expectations of war were formed by World War Two or the Korean War, is the Ling-Temco-Vought A-7, a light attack aircraft used by the navy for visual conditions only (though the A-7D, built for the air force, has all-weather radar bombing ability). Its design was based on that of the Chance Vought F-8 Crusader as an economy measure (the aerodynamic characteristics of the arrangement being already well known), and it looks like an embryonic version of the Crusader with more infantile, less ferocious features. Subsonic, it substitutes rounded leading edges on wings, airscoops, and radome for the Crusader's sharpened wedges; it is shorter, chubbier, has a proportionately larger wing, and lacks the F-8's unusual provision

A-7 Corsair II is a single-seat, single-engine light attack plane that replaced the A-4 Skyhawk.

59

for altering the pitch angle of the fuselage with respect to the wing.

Popularly called the Corsair II in honor of the bent-wing "hose-nose" of World War Two, the A-7 is the smallest airplane on the carrier, though larger than the Douglas A-4 which it replaced. With a wingspan of under thirty-nine feet, barely larger than that of a four-seat single-engine light airplane, it can carry over 15,000 pounds of external ordnance—bombs, air-to-air and air-to-ground missiles, rockets, or pod-type guns. Like almost all other navy carrier aircraft, the A-7 has a retractable refuelling probe, this one mounted beside and below the cockpit on the right side of the airplane. On the left side of the A-7's nose is a 20mm cannon

capable of firing thousands of rounds per minute, but able to carry only about ten seconds' worth of ammunition. Although the A-7 can carry heat-seeking air-to-air missiles, its main role is close air support and attack, meaning that as ground troops advance to seize the enemy's territories, the A-7 swoops low overhead, bombing enemy gun or missile emplacements, strafing vehicles with rockets and gunfire, and tossing fire- and fragmentation-bombs upon enemy soldiers.

F-18 Hornet of VFA-303 rolls elegantly away from Air Force KC-10 tanker after refueling.

60

The A-7 leaves the carrier at about 40,000 pounds gross weight; its fanjet engine develops about 12,000 pounds of thrust, giving it roughly the same thrust-to-weight ratio and performance as the A-6.

The newest tactical aircraft on the boat, and an aircraft clearly destined to have a profound effect on carrier operations, is the McDonnell Douglas–Northrop F/A-18 Hornet. As its designation implies, the F/A-18 is a dual-role aircraft, capable of air combat and interception as well as ground attack. The Pentagon's track record with multi-role weapons has been dismal, but the Hornet has distinguished itself in a short time as a spectacular exception.

The F/A-18 will eventually replace the A-7 as the carrier's light-attack bird, and several ships have already made the switch. The future mix of aircraft types on carrier decks will undoubtedly shift, and the Hornet will be at the center of these changes. It is capable, without any alteration—indeed with nothing more than the flick of a switch on the control column—of going from ground attack mode to ACM (air combat maneuvering, or dogfighting). The same Hughes digital radar is used for both ground-attack and for long-range air-to-air intercepts; in the latter mission the Hornet can carry AIM-7 Sparrow and AIM-9 Sidewinder missiles, plus the newer AMRAAM, which is still in development. The ultra-long-range Phoenix remains the exclusive province of the F-14.

The F/A-18 is a single-seat, twin engine, twin-tail fighter/bomber, considerably smaller than the huge Tomcat but larger than the A-7. Its General Electric F-404 afterburning engines, putting out 16,000 pounds of thrust each, give it a thrust-to-

weight ratio better than 1:1 at half fuel; it can turn and burn with any fighter on the planet. All flight controls are fly-by-wire; the airplane's bewildering collection of control surfaces move in computerized concert, sometimes eight or ten different surfaces changing position in one maneuver. It's a strange and wondrous spectacle for those used to aerodynamically simpler aircraft, but the Hornet has proven in its few years of service to be as reliable and problem-free as any plane in the navy or marine corps.

The F/A-18 is amazingly easy and comfortable to fly, with its unprecedented degree of automation and its well designed cockpit and displays. It is particularly pleasant to land on the carrier, even without the ACLS, because it gives the pilot magnificent control and stability. Carriers with F-18 squadrons usually find the Hornets leading the race for best landing grades. In almost 200,000 hours of flying, the U.S. Hornet fleet has turned in an extremely low Class A accident rate of 5.2 per 100,000 flight hours—one fourth the rate of the F-14 in its early years.

The Hornet has proven almost too good to be true—a true dual-role aircraft with breathtaking performance as both fighter and mud-mover. Its one drawback has been excessive fuel consumption, producing shorter strike range than hoped for. Engineers are working on squeezing another two to three thousand pounds of fuel into the airplane to rectify this drawback.

A beefed-up two-seat F/A-18D is down the road; this aircraft will serve as an all-weather strike bomber that can also engage enemy fighters when it pulls off the target; photo-recon, battlefield-observation, radar-jamming, and "Wild Weasel" (SAM

61

suppression) versions can be expected in the future.

The antisubmarine role, formerly performed by a version of the Grumman S-2 twin-radial engined airplane, has been taken over by the Lockheed S-3 Viking, a shortcoupled, attractively proportioned airplane with two 9,000-pound thrust high-bypass turbofans slung in pods beneath the wings, quite close to the fuselage. The high-bypass engines are of a distinctively large cross-section and produce a resonant and penetrating hum unlike the sound of any other carrier airplane. Light in weight and with a large wingspan and fuel-efficient fan engines, the S-3 can range 2,000 nautical miles on its internal fuel load of less than 2,000 gallons. It has an on-off, down-only type of DLC.

The S-3 has a crew of four. The pilot commands the aircraft; the copilot, in addition to sharing flying duties, navigates and manages radar and infrared target sensing equipment. A third crewman, called the "Senso," interprets information transmitted back by data link from acoustic-sensing sonobuoys that are dropped into the water from the airplane (which carries sixty of them). Information is stored in an onboard computer, which also computes aim points, weapons trajectories, and navigational plots. A fourth crewman, the Tactical Coordinator or "Tacco," manages the attack on a submarine once it has been detected. In addition to acoustic sensors, MAD (Magnetic Anomaly Detection) equipment mounted in a retractable boom on the tail of the airplane can detect the mass of ferrous material represented by the submarine at some distance beneath the surface. Offensive armament includes bombs, mines, destructors, depth charges, rockets, and homing torpedoes.

The current radar patrol aircraft is the turboprop Grumman E-2 Hawkeye. Its role is to loiter at a distance from the carrier and to return to the Combat Information Center aboard the ship electronic data collected by its radars, of which the most conspicuous is contained in a twenty-four-foot rotating disc supported on a pylon above the airplane. Information is returned to CIC by data link and can be displayed on several scopes in several different forms, at the choice of the operators. Electronic sensors and a powerful computer are able to single

S-3 Viking provides the carrier with long-range anti-sub protection. Its Magnetic Anomaly Detection (MAD) booms, seen extended, can detect the sub's metallic mass far beneath the surface.

62

out and identify targets against heavy background clutter and supply continuous information on speed, bearing, altitude, and heading of aircraft. The purpose of the E-2 is not only to provide early warning of approaching enemy ships or aircraft, but also to direct the carrier's aircraft to interceptions. The carrier surrounded by its patrol aircraft proceeds less like a single battleship, or even a task force, than like a swarm of bees, with a dense core of concentrated force and a fluid shell of constantly moving outriders, the whole ensemble reaching out over an area hundreds of miles in diameter.

In addition to fixed-wing aircraft, carriers use a complement of helicopters, Sikorsky SH-3 Sea Kings, for antisubmarine work and for water rescue during launches and recoveries. The Sea King is big: 20,000-pound gross weight, 62-foot main rotor diameter. Its antisubmarine equipment in-

cludes sonobuoys, sonar, MAD equipment, and radar.

The Sikorsky SH-60 Sea Hawk, a navalized version of the army's Blackhawk troop carrier, will soon replace the aging SH-3 in the SAR and close-in ASW role. The Sea Hawk is already at sea with the navy, flying from surface ships as the airborne element of the LAMPS (Light Airborne Multi-Purpose System) antisub technology.

The mix of aircraft on the carrier deck is constantly changing; one of the greatest strengths of the big flattop is that it can modify its capabilities in only the time it takes to fly one type of aircraft on and another off. The 1990s will see more and more F/A-18s in both single- and two-seat configurations, the new A-6F Intruder with up-

E-2 Hawkeye with F-14 escort. The Hawkeye provides detection and analysis of airborne threats at extreme range.

63

to-date electronics and the F-404 engines used in the Hornet, beefed-up S-3B Vikings configured to carry Harpoon antiship missiles, and the vastly more capable F-14D. Design work is underway on the Advanced Tactical Fighter (ATF) and an Advanced Tactical Aircraft (ATA, a long-range all-weather attack craft) that may replace the Tomcat and Intruder at the end of the century.

Many prognosticators have forseen V/STOL aircraft as the carrier planes of the future; such aircraft, say their proponents, could easily operate from smaller and cheaper carriers. The "small carrier" argument hinges on V/STOL, and as yet the technology has not produced an aircraft that can perform all the required missions. The navy and marines do in fact operate small carriers with V/STOL capabilities; these are the eleven LHA and LPH helicopter assault ships. Aside from helicopters the

only operational V/STOL jet is the AV-8 Harrier, a light attack bird of interesting but decidedly limited talents. The latest version, the AV-8B Harrier II, is a far superior airplane to the original, with greater range, ordnance payload, flight stability, and dependability. In fact the Harrier II compares favorably with other light attack birds such as the A-7. But it is strictly subsonic, leaving it unable to compete with serious fourth-generation fighters such as the F-18. More impressive V/STOL fighters and attack craft will undoubtedly be developed in the future, and the entire raison d'etre of the large-deck carrier may shift as a result. But in the meantime there remains no substitute for the phenomenal capabilities of the large-deck aircraft carrier and its air wing.

SH-3 Sea Kings returning to Kitty Hawk. SH-3 "helos" provide the ship with air-sea rescue and close-in anti-sub capability.

64

McDonnell Douglas AV-8B Harrier II, the latest version of the amazing English jump-jet, is operated by Marine squadrons on land and aboard Navy LHA and LPH assault carriers.

...AND THE MAN

Tomcat of VF-21 Freelancers lines up for trap aboard Constellation (CV-64).

THE LIFE of carrier pilots is by turns sporting and monastic. They candidly regard their airplanes as being similar to toys or to the luxuries of the rich, like very expensive cars; and their lives revolve around the few hours a day they spend flying or preparing to fly. They live below decks in tiny staterooms, usually two to a room. The ceilings are threaded with pipes, cables, wires, hoses, and ventilating ducts. A small stainless-steel sink is surmounted by a mirror whose metal frame is painted pale green. There is a locker for each man, a small writing desk, some drawers and shelves. A chair. Steel bunks with thin mattresses and grey woolen blankets have at their head small incandescent reading lamps. There are no portholes; but a framed picture of a woman is a window through both space and time to another world from which this one is remote, snugly insulated, and almost inaccessible. That woman—or mother, or child—will never know the inside of this Spartan cell, nor will she ever construct in her imagination a precise conception of the qualities of this austere life which make the flyer leave her for it.

They fall asleep to the lullaby of ventilators and awaken to the clatter of cables or the sudden whirr and thud of some hidden motor. They eat institutional meals in cafeteria-style wardrooms, and spend leisure hours sprawling in the leatherette chairs of the ready room exchanging jokes and jibes, reading magazines, or studying navy manuals. At night if they are not still flying they sometimes crowd into a room to watch movies that flicker upon a makeshift screen; between reels they drink cups of Kool-Aid.

If seagoing pilots are bored or lonely, they do not show it. All of them have some collateral duty during the hours they are not flying, so they work for twelve hours a day and have little empty time. The camaraderie of the ready room and the sky, simplifying their emotional lives to a few powerful and harmonious strains, heals rapidly whatever spiritual scar tissue might be left by the sporadic quality of family life. With their copilots, RIOs, and fellow squadron members, they find the periods of mental synchrony, and an almost telepathic understanding, which banish loneliness.

68

F-14 jockeys from
Fighter Squadron 111
suit up aboard Kitty
Hawk.

Each day brings a schedule of events. Life is a permanent rehearsal. Half an hour before a launch the pilots leave the ready room for a small dressing room down the passage; there they suit up. Their olive-drab flying suits look like bodies turned inside out, with all the viscera on the outside. Air ducts in elastic passages, which are hooked up to a system in the aircraft when they climb into the cockpit, provide compressed air to the G-suit in response to G-load, constricting around the lower part of the body and keeping blood from draining out of the head under heavy accelerations. The suit is a mass of zippered pockets in which survival equipment is kept. Over the flight suit is worn a kind of parachute harness, sans chute, that will be hooked to the ejection seat in the cockpit. Under the flight suit the pilots wear the turtleneck jersey of the color of their squadron.

It is when they saunter across the swirling deck in their flight suits, kneepad and helmet in hand, in casual pairs or files single or grouped, that their grand moment comes, their knighthood flowers. They are now the athletes taking the field, the stars for whom all the stagehands sweat; they are engulfed in the roar of an imaginary crowd. Their airplanes, titanic and tame, stand arrayed, canopies open, ladders deployed, fuelled; armed, ready. Calm, without formality, the pilots and the RIOs check the planes over, climb in, buckle themselves to harness and seat and, elbows akimbo on the cockpit rails, sit peacefully, sublimely, waiting for the games to begin.

It costs the navy a million dollars or so to train each of these pilots for this job. Consequently he commits himself to years of flying once his training is over; for some it will be a career. He will have gone to Pensacola for his basic officer and primary flight training, then gotten his wings, qualified for carriers with a number of arrested landings on shore runways and then on a carrier, and learned the basics of air-to-air and air-to-ground gunnery, bombardment, dogfighting, radio navigation, instrument flying, formation flying, and so on. By the time he gets his wings he will have 250 to 300 hours of flying time. Then he goes to a RAG—Replacement Air Group—to specialize in the airplane to which he has been assigned.

69

In the RAG he flies another 100 hours or so, goes to ground school to learn the airplane's characteristics and systems in detail, and repeats, in advanced form, much of the warfare practice he has already had.

At the end of two years and 400 hours he will be in a fleet squadron flying on carriers (if, that is, he does not for one reason or another go down a different pipeline into one of the many ground-based kinds of navy flying: transport, recon, patrol, training, or target towing).

Henceforth his career will alternate between ship and shore assignments of about two years. While he is on sea duty, his ship will probably deploy on a cruise of six months or so; during this time he may not see his family, his home, or friends other than his fellow pilots. In exchange for these deprivations he receives a sense of accomplishment, the gratification of his patriotic impulses, a chance to see the world, and above all, the opportunity to be part of the grand sport of carrier flying.

Squadrons are predominantly composed of young men—newly out of college, ensigns, JGs, lieutenants—because many pilots do not remain in the navy after their first tour of duty. There is more money on the outside, and fewer demands on one's private life. Many go to airline jobs, or to seats in the cockpits of corporate jets or charter airplanes. They know that they are well trained and in demand. Furthermore, many feel that the heyday of their flying career will have been their first tour; only the best of them will still be able to get a desirable, sea-going flying billet after a tour of shore duty. Those who stay in and distinguish themselves as pilots and leaders find their ways to positions as lieutenant command-

ers and commanders: XO or skipper of a squadron, commander of an air group. A few make captain and are still flying; a very few eventually will get a command of a carrier—the captain of a carrier is always a flying officer.

But the population tapers off sharply with age; the majority of navy pilots are, like the majority of athletes, men in their twenties and early thirties. The tone of their conversation in the ready room or wardroom, almost always irreverent, bantering, mutually insulting, off-color, rowdy, is the tone of fraternities and locker rooms. Often a pilot has a nickname chosen by his mates, mildly insulting if only because it is rarely the one he would have chosen for himself. The pilots are a different sort of person from the other ship's officers: more casual, independent, indifferent in hierarchy, dignity and tradition, as though they do not need the navy as much as the navy needs them. Whereas it is an indispensable piece of protocol among navy men that all large surface vessels are called ships, flying officers, including carrier captains, make a point of calling the carrier the "boat." The message is not lost on the ship's company.

Not much flying is done when the ship is cruising from one port to another. The peacetime navy avoids "blue water ops," carrier operations out of reach of a land "bingo base" to which a plane can return if for some reason—most often mechanical—it cannot land on the carrier. If something goes wrong during blue water ops, there is always the barrier net; but it damages airplanes and is used only as a last resort. So the remote transits are boring times. Otherwise pilots fly once or twice a day, in "cyclic ops" with a round of

70

launches and recoveries every hour or two. They are always practicing: radar intercepts, air combat (in which navy pilots excel), strafing a "spar" towed behind the ship, bombing land targets, shooting at air-towed "darts," antisub operations—and landing, landing, landing.

On the cat shot off the deck, the pilots like to say, you're along for the ride: just a passenger. The airplane is trimmed for the climbout. The cat rockets you off at maximum power or "in burner," five or six Gs for a second or two, and then a floating feeling as normal flight takes over. Night cat shots are a little scary: nothing to see but the red glow of the cockpit, a few peripheral lights, perhaps a few clouds outlined by the moon, some stars; or under an overcast, only a velvety and tomblike blackness. You brace your head against the seat, perhaps jam an arm up against the glare shield, and ride up into the dark. You're ready to punch out in an instant if something goes wrong: a disabled plane coming off the cat does not take long to hit the water. No delay is possible. An A-7 pilot had a gear strut lose pressure as the cat fired at night; he felt the

Corny "call signs" relentlessly follow pilots through their careers.

airplane tilt. The Air Boss had seen it happen before, seen a pilot killed: he shouted "Eject!" and without stopping to think the pilot punched out. The plane hit the water a moment later, and so did he. Television tapes told the story: from the cat shot to the ejection, two and a half seconds had elapsed.

When you have flying speed you clean up: gear, flaps, burner off, switch radio frequency, off to the races.

The hard part is getting back aboard.

Visual carrier landings in good weather are not terribly difficult once you have the hang of them. A good land pilot could come aboard reliably in a COD with a week or so of practice. The pattern is flown tight, passing by the ship downwind at 800 feet or so, dirtying up in the crosswind turn, turning —one continuous turn, no base leg—at approach speed abeam the fantail and rolling out on final a third of a mile or so behind

71

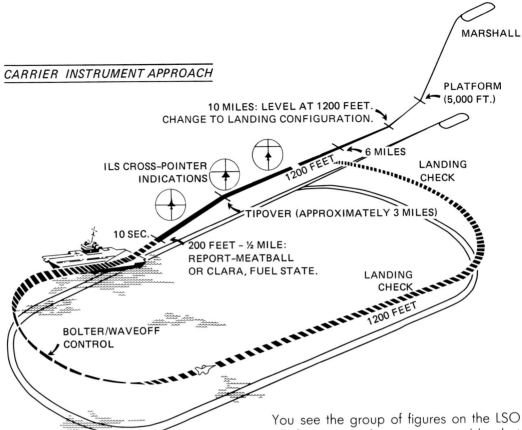

CARRIER INSTRUMENT APPROACH

MARSHALL

PLATFORM
(5,000 FT.)

10 MILES: LEVEL AT 1200 FEET.
CHANGE TO LANDING CONFIGURATION.

6 MILES

LANDING
CHECK

ILS CROSS-POINTER
INDICATIONS

1200 FEET

TIPOVER (APPROXIMATELY 3 MILES)

10 SEC.

200 FEET – ½ MILE:
REPORT-MEATBALL
OR CLARA, FUEL STATE.

LANDING
CHECK

1200 FEET

BOLTER/WAVEOFF
CONTROL

the ship. There are certain things to look out for. Because height of the optical glideslope is adjusted by rolling the plane of the glidepath along an axis aligned with the edge of the deck, the glidepath is often low to the left of the ship. You have to ignore the ball until the very end of the turn to final; but you also have to ignore visual cues in general and fly instruments.

You should come out right under the glidepath, intercept it immediately, and start down on the ball. You don't have long to correct your lineup; but the deck, clearly in view right in front of you, is easy to line up on. The aim point is the crotch of the deck, the forward edge of the angled deck where it meets the axial deck. It expands rapidly.

You see the group of figures on the LSO platform, see the arresting cable that trapped the last airplane scurrying back across the deck to meet you, your scan marches rapidly through angle of attack, ball, lineup, the deck suddenly storms toward you, explodes into the periphery of your vision, you slam down as the throttles go full forward, the plane yanks you backward as though it had struck a wall of muddy sand, and you're aboard.

Under good VFR (Visual Flight Rules) conditions, in daylight on a smooth sea, the most difficult part of the carrier landing is the selective perception it requires. In close, a pilot who fails to limit his perceptions is said to be "spotting the deck"; he gazes at the deck. This may seem a minor sin—even a virtue—but experience has shown that

72

safety is better served if the pilot keeps his scan moving between the meatball, the centerline, and the simplified cockpit angle of attack indicator with its chevrons and doughnuts than if he looks fixedly, as a land pilot does, at the approaching runway. Day or night, if you look at the deck you always feel you are high and want to drop down; but nothing is more dangerous than arriving even a few feet low. The pilot who spots the deck is inviting his erroneous visual impression of a high approach to undermine his concentration on the ball. The result is "settling in close"—getting too low in the last eighth of a mile.

Another cause of settling in the final moments of the approach is the disturbed airflow behind the carrier. Especially when it is calm at sea and the wind over the deck is produced by the carrier's motion, there is a crosswind on the angled deck and the wake of the island crosses the approach path close behind the ship. When the wind is straight down the angled deck, less turbulence enters the glidepath, and it is farther behind the ship; but it is still there. The wind tumbling into the wake of the ship after passing over the deck also provides a downward component at the ramp, regardless of the wind direction.

Finally, it has been found that pilots in general tend to fly the approach a little high in daytime and a little low at night. Arriving a little high leads the pilot to "dive for the deck" at the last moment, responding to a high ball. But the meatball is deceiving in close and at the ramp; the depth of each ball over the ramp is only about thirty inches. Obviously it is impossible, in the last one or two seconds of the approach, to make an accurate adjustment of two feet in the height of the airplane and stop the adjustment on the money; instead the pilot is likely to drop through the proper height and hit the deck hard, or catch a one wire on the ramp, without ever being aware that he went low.

Instrument approaches of the simpler kind are not much different. In a Case Two approach, with better than a thousand-foot ceiling, you come straight in and may pick up the ship from ten miles out; thereafter you go over to tower and make a VFR approach. It's here that you would notice, more than in a circling VFR approach, the slight crab angle, or the continual succession of sidestep maneuvers, that are necessary to keep lined up with the angled deck.

Because the deck is at approximately a ten-degree angle to the direction of motion of the ship, the ILS course in effect moves sideways at about five knots, requiring a heading correction of about two degrees to the right in addition to any other corrections required because of wind, instrument errors, and so on. During the ILS portion of the approach the corrections all blend together, since you cannot see the ship; when you break out, you may be dimly aware of the small change in heading just before hitting the deck, though from farther out the correction for ship's movement still feels like a crosswind correction.

Case Three approaches are conducted, according to the book, down to 200 feet and half a mile. In wartime this protocol cannot always be observed; there may be no bingo field, and you land either on the carrier or in the water. When recoveries must take place below minimums and the Automatic Carrier Landing System is used, you are again, as on the cat, a passenger.

Hands riding lightly on the controls, which stir of their own volition, you stay ready to take over if something goes wrong. You can't always be certain that everything is under control; a sudden sidestep or pullup, meaning that the radar has lost lock on the transponder or reflector and picked up the falling rain or the wingtip, can shake your confidence marvellously. It's an eerie feeling: the controls moving, the throttle creeping back and forth, and the illusion of a thinking creature directing the immediate and appropriate responses of each control to the deflections of the ILS needles.

Hand-flown, the carrier approach to minimums is simply an ILS approach that must be flown very accurately. As with all other approaches, the characteristics of the airplane are the most important factor, apart from the pilot's acumen, in the difficulty of the approach. The latest airplanes are not necessarily the best. With its short, highly swept wing, high wing loading, and powerful turbojet engines, the F-4 Phantom was almost impervious to gusts, and its thrust line was so well oriented with respect to its center of gravity that power changes did nothing to alter the pitch attitude of the airplane. Power provided direct control of height with respect to the glideslope, and the power response from the modern turbojets was prompt.

The F-8 Crusader, on the other hand, had some dreadful handling characteristics, although it is still regarded throughout the fleet as a "tits machine" now that it is no more (the French still operate them on carriers). Its landing speed was very high— 155 knots—faster by a fifth than anything landing on the boat today. Landing gear collapses were far from unheard of. Its tur-

bojet engine, situated in the rear of the fuselage, was tilted slightly upward so that when the two-position wing was raised to its landing-incidence position, the fuselage was horizontal and the engine sat slightly nose-down (this odd state of affairs kept the main landing gear short on a rather long airplane). When the Crusader pilot added power on the approach, his airplane wanted to nose downward, pushed that way by engine thrust. By current standards the F-8 was an unsafe carrier airplane; of sixty-nine serious or Class A carrier accidents in the sixties and seventies, half involved F-8s.

The more efficient, still more modern fanjet engines are again slow to respond to the power lever and most lack the F-4's good pitch stability. In the A-7, for instance, the power is a second or two late in coming, and when it does come in the airplane undergoes a small pitch trim change. A pitch change with power isn't too disconcerting when it arrives simultaneously with throttle movement; you get into the habit of automatically moving the stick at the same time as the throttle, or only a split second later. When power response is slow, however, you have to insert a delay proportionate to the amount of power change or listen to the engine to know when the power is coming in.

The civilian pilot on an ILS approach must have the needles centered when he reaches the Missed Approach Point, or minimums; but if he then breaks out, he can continue the landing visually, and precision ceases to be necessary. For the carrier pilot, on the other hand, precision is required right down to the deck. It is almost impossible to fly a "perfect" approach; the test of the pilot's

74

Korea, 1953: a burning Banshee is inundated with foam aboard Lake Champlain.

skill is his ability to correct deviations promptly and accurately, and without overshooting into another deviation.

A mile behind the ship, the rates of response by pilots to instrument indications and by airplanes to control input are well within the rate at which deviations from the glidepath occur; but as the airplane approaches the ramp the rate of transition from, say, a center ball to a flashing red low ball becomes more and more rapid. In close, certain corrections are impossible; they won't be completed in time, or they are bound to lead to an error of another sort. A high ball in close, for instance, should not be corrected; better to hold the high ball down to the deck and take a four wire, or bolter, then try again. The worst reaction would be to correct with the nose.

Pilots eventually learn to anticipate the results of their corrections so that they are not always busy catching up with their own mistakes. They also learn what types of corrections are appropriate at which portions of the approach, and which trends are likely to lead to trouble.

The most difficult and dangerous years of carrier flying were during the Korean War, before carriers were equipped with precision instrument approach facilities. Regardless of weather or time of day, all approaches were made in the classic military pattern: overhead pass at low altitude, break, dirty up, short turn to final, and almost immediate landing. The task of landing the early jet fighters—with their high wing loadings, high approach speeds, and slow throttle responses—on a straight-deck carrier in darkness, bad weather, or both, was at the very margin of human abilities.

An F-8 pilot ejects from his crippled bird over the South China Sea in 1965 as his wingman looks on. The small drogue parachute is about to pull out the main chute. This spectacular photo was taken by the side-looking nose camera in an RF-8A reconnaissance aircraft.

Fatigue, a moment of fear, a slight misjudgment, vertigo, or any slight disturbance could put a fighter over the wires and into the barrier or the parked planes beyond. Some squadrons suffered an attrition rate of ten percent a year.

Today, landing on the carrier in daylight is not nearly so difficult, even in bad weather.

If the navy has sixty-hour pilots doing it, one can imagine that the task is not out of the reach of any experienced and reasonably adept pilot. Even without the help of all the modern electronic landing aids and backed up only by the experience of the LSO, the skilled pilot makes one daytime trap after another without mishap. Ninety percent of day VFR approaches end in traps.

Landing on the carrier is similar to all tasks of precision flying: the pilot must quickly and accurately perceive his position, his motions, and rates of change of motion, and integrate them according to his experience to put him where he wants to be. In daytime he has a supply of visual clues that proceed to his fingertips instantly, without any intervention of conscious thought, and are translated into maneuvers.

But at night most visual cues are abolished. Unless a full moon is shining over a cloudless sea, the only lights come from the carrier: the runway lights, the red light atop the island, a glow surrounding the island, and the ghostly, gossamer haze that emanates from the hangar deck and settles upon the passing water.

The night landing, especially in rainy weather, makes the worst psychological demands on the pilot. Even a daytime approach to minimums ends with a final few seconds of visual contact with the carrier, a chance for a little body English to refine the lineup, and a bracing taste of reality. But on a very dark night, reality begins only after the trap: up to the instant of hitting the deck, you seem to be flying in outer space. Small rotary motions of the body, like those produced when rolling out of the final turn, can produce spatial disorientation and vertigo. Pilots of the fifties who marshalled overhead

76

and approached visually tell of horrifying illusions: as they turned to final, the carrier, its lights, and the glow of its decks all seemed to rear up out of the water and hang vertically before them like a breaching whale.

It is difficult to imagine the sensations of the night approach. If you place yourself in the darkest subterranean room, look across it at two or three points of colored light and a vague, faint speckling of others, and then imagine hurtling toward that dimensionless

From the standpoint of preventing vertigo, a straight-in approach would be best; but from a few miles out the ship appears as no more than a point of light in the blackness. So the night approach is always flown down the ILS.

The most modern version of the ILS receiver is found in the F-14, which uses a cathode ray tube to display the horizon, the ILS crosspointers, the heading, and even a "fly to" cursor indicating the heading to fly for proper wind correction. Starting the ILS

puzzle at 150 miles per hour, threading the needle between the ramp and the four wire while feeling all the while that you are standing still, you may begin to get some idea of the approach. But you cannot simulate the visual disorientation that hovers at the edge of the pilot's awareness; we are never aware of this sensation when there are plenty of visual cues around us and our brains are constantly checking information from the kinesthetic organs of the inner ear against that from the eyes.

the pilot sees only a point of light in the distance ahead of him; this he knows is the carrier. He flies the needle, gets the airplane settled down on the right heading, the right power, the right angle of attack. In the F-4 it was easy; once you had things set up, nothing changed much. In the A-7 there is always a lot of throttle-jockeying, because of the delays in engine spoolup. In the F-14 the throttle takes care of itself. The pilots fly good approaches; if you watch them in the simulator, practicing, you can see the dis-

77

play they are seeing: they get the needles crossed and centered and then make only very small corrections, promptly, so that only the smallest deviations occur. Things get sloppier if the simulator operator throws in, for instance, an engine fire warning light; but even with this distraction, their ILS performance is very good.

The accuracy of approaches can also be appreciated on the television monitors found all over the carrier. A camera sights up along the flightpath through a hole in the flight deck; crosshairs indicate the center of the glidepath. The television image reveals every deviation from a perfect approach; and with remarkable regularity, one pilot after another remains in the crosshairs for the entire approach or, if he strays from them, returns promptly and positively.

At a mile or so, in the groove, the pilot looks up and clearly sees the ship. It appears as nothing more than a slightly tapered box of white lights, with a centerline that extends out of its bottom edge, turning to orange: this is the dropline hanging down the stern of the ship. To the left of the box, a little past its middle, is the "meatball"—the optical glidepath indicator. It's surprisingly small; a horizontal row of tiny green lights with a single orange light at the center. To the right, and apparently slightly above the box, is the red light atop the island, and sometimes a cluster of white lights emanating at random from the island. Sometimes he sees the glow on the water, which tends to place the pattern of light into perspective; but if he cannot see the water, the pilot can deduce the perspective of the deck only because he knows in advance what he is going to see. The meaning of the lights is abstract, it is unpersuasive; pilots have

landed in the water behind the ship, because they somehow imagined that they were on the proper approach path and realized their mistake only when they saw the reflection of their own lights on the waves.

The box looks more or less square. If it were a land runway its shape would indicate that you are very low—much lower than the three-degree glideslope would suggest. But the spacing of the lights changes the impression; the more distant lights do not seem much closer together than the nearer ones, and so you clearly perceive that it is a very short deck. In fact, it appears as though you are flying into a square hole almost directly below you. Paradoxically, the square runway outline makes you feel high, not low; and there is at night a tendency to go low—against which the only protections are the blinking red low ball and the judgment of the LSO. To avoid going low, pilots may try to approach slightly high at night, and four wires and bolters are far more common at night than during the day. In fact, at night in bad weather, the recovery rate drops to thirty percent.

At a mile out, the meatball is not only tiny but indistinct, day or night, and its movements can be very hard to judge. Once the pilot has stopped flying the ILS needles and called the ball, he relies heavily on the LSO for good height information until he is in closer. He can't chase the glidepath; he holds his rate of descent at about 500–600 feet per minute and believes first in the LSO, only second in the meatball. A mile out, one cell of the LLD is thirty-two feet deep; the LSO's judgment is consequently more accurate than the pilot's.

78

An F2H breaks up after striking the ramp aboard Oriskany off Korea in 1950. The pilot ejected safely.

From half a mile out the outline of lights upon ghostly water may begin to look like a ship. You are approaching it rapidly—about thirty seconds to go—but it does not yet seem to be expanding toward you at a dizzying rate. This is the phase of the approach that most resembles a day approach: you have a good visual fix on the ship, and even without peripheral cues you can deduce from the meatball and the dropline when you are on centerline and on the glidepath. Your instrument indications are projected in space before you by the HUD, Head Up Display. You feel as though you could continue the approach to a landing this way without other cues. But the last few seconds are different.

As you get close to the ship—"at the ramp," as the pilots call the last stretch—the impression that the lights outline a hole becomes stronger. You feel as though you are flying through a hoop. The stable sense of position and aim point that you had thirty seconds earlier suddenly seems to explode into fragments as the box of lights expands rapidly. In the last few seconds there are only two cues: the centerline lights and the ball. The essential thing is the ball; by this time lineup should be pretty good—the approach would have been waved off if it were not—and the great danger is that of hitting the ramp.

An airplane that hits the ramp usually explodes instantly into flames, spewing

79

burning wreckage along the deck. The crowd on the LSO platform abruptly vanishes (the trick is neither to jump first, so that everyone else lands on top of you, nor last, when there is no room left in the nets, but at some intermediate point assuring both your comfort and your safety). The pilot ejects at the moment the airplane strikes; as the explosion clears, his parachute is seen settling into the water in the wake of the boat. If he does not eject then, there is still a tiny chance he will survive: the severed cockpit capsule of the demolished airplane could slide to a halt on the deck, with the unscathed pilot still inside.

Though hitting the ramp is the worst outcome, landing to the side of the centerline could be dangerous as well. Sufficiently far to the left, a landing gear would go off the edge of the deck; the airplane would fall into the nets and, in all likelihood, "go over the side" and into the water. Navy films show an F-8 Crusader pilot going over the edge this way. The incident was memorable; the plane hung in the nets as flames began to lick around it. The pilot, who had hoped up to the last moment that he would not go over the edge and had failed to eject until it was too late, clambered out of his cockpit. Despite the flames and the narrow escape, he was not in a hurry to leave the plane. He stood on its side while deckhands reached out to him. With a gesture sublime in its relaxed gallantry he removed his helmet and handed it to one of the deckhands, precisely as a dinner guest at a state banquet might once have laid his gloves across his top hat and handed it to a butler. He then accepted another outstretched hand and stepped over onto the deck.

Another cause of accidents is the air-borne engagement. Feeling at the last moment that he is low, or receiving a wave-off and reacting late, the pilot pulls up the nose of the aircraft and adds power at the ramp. Pitching the nose up lowers the tailhook, however, and it is possible for the hook to engage a wire without the wheels touching the deck. The wire will stop the airplane regardless; and the airplane pancakes down on the deck, usually collapsing at least the nose gear and possibly all three.

In a related but less drastic sequence of events, the hook may strike the ramp and bounce over all four wires. The hook is equipped with a hydraulic shock absorber to prevent its bouncing, but it is capable of bouncing nevertheless; sometimes an acceptable pass ends in a bolter for this reason. It is because a trap is never certain, no matter how good the approach, that full power is always brought in as the airplane crosses the ramp.

Problems with height over the ramp are most common when the ship is pitching or heaving strongly; the stabilization of the LLD goes only so far, and in heavy seas the glidepath is capable of waving up and down behind the ship. Studies have shown that the response curve of the pilots and engines and the pitching rate of the carrier interact in such a way as to put the airplane 180 degrees out of phase with the ship: when the glidepath is starting upward, the airplane is settling downward, and vice versa. Under those conditions getting proper height over the ramp is difficult.

Tomcat is positioned for cat shot aboard Ranger (CV-61).

80

The one element that simplifies the carrier approach is the absence of a flare—the transition from a descent into level flight just above the runway. It is not necessary for the pilot at the last moment to make that complex judgment of speed, height, and pitch rate which is necessary for ordinary landings of civil aircraft. The civilian pilot, who is concerned with landing more or less smoothly, finds the idea of landing a jet consistently within a 120-foot stretch of deck astonishing, because he remembers his own difficulties with putting the wheels exactly where he wants them. But the carrier pilot does not have to cope with the complication of the flare, with its sudden multiplication of variables and the unintended conversion of a knot or two of speed into a hundred feet of runway. The carrier pilot needs only to fly a straight line. His most difficult moment is the last moment, but not for the same reason as for a civilian pilot. For the carrier pilot it is simply that the last hundred feet of the approach bring him into the narrowest constriction of a funnel which was, farther out, more tolerant of error. If he had been weaving above and below the glidepath or from side to side a mile out, his corrections would not have put him outside the parameters of a safe approach. At the ramp, the same corrections will be excessive and perhaps dangerous.

Landing at night on a carrier has something in common with driving at high speed through a narrow gate. At a distance one can aim; close in, corrections become intuitive and too quick to be judged consciously; at the last instant, control is barely possible as the target rushes up to meet you. It is that last instant that divides success from failure, perhaps life from death.

Every carrier pilot talks about the "pucker factor" of the night landings. No one dismisses this or claims to have it well under control. During the Vietnam War the navy instrumented some of its pilots to collect data on physiological indications of stress—heartbeat, breathing rate, and so on. It was found—and this result has been enshrined in perpetuity in the annals of carrier lore—that the stress levels were never so high, not even when dodging SAMs and MiGs over Hanoi, as when approaching the carrier at night. On average, a navy pilot makes about one-third of his landings at night; there is no letup, no relief. There can be no inaccuracy, no neglect, no inattention. More than anything else, it is the night landing that gives carrier pilots their pride, makes them claim that there are no other pilots like them, and serves as the rite of passage to their knighthood. It is also probably the only part of their peacetime flying that they do not enjoy.

E-2 Hawkeye makes an "OK" pass aboard Constellation.

82

PAST AND FUTURE

Eugene B. Ely, history's first carrier pilot, touches down on U.S.S. Pennsylvania in 1911. Ropes tied to sand bags were used as arresting gear.

THE HISTORY of the aircraft carrier is the history of a time of rapid change in naval warfare. Carriers were born in the era of capital ships—battleships and battle cruisers—whose roles were those of blockade, bombardment, and sea warfare. The scale of all warfare was different then; distance was still a protection. A fleet or an army not wishing to engage another could slip past it; the power of ships did not extend beyond the horizon. Ships were armed with guns which, though they might range twenty miles or more, still were paltry weapons compared with airplanes.

In the past thirty years, however, the navy has waged war against its own obsolescence. A few of the traditional roles of classical sea power, such as control of sea lanes for shipping, still exist. The navy can still mount blockades, as in Cuba in 1962, or break them. In Vietnam ships were used to bombard the shore with guns, and some were hit in turn by shore batteries—an outmoded style of warfare, but still a feasible one if the possessor of the shore lacks an air force. But whereas a century ago fleets clashed on the sea as armies clashed on land, and even during the Second World War battleships and cruisers engaged one another with guns, the main role of the navy has gradually changed.

By adopting submarine-launched ballistic missiles and ship-based tactical aircraft, the navy has coopted some of the tools of the army and air force and adapted them to the marine environment. Ships in themselves are no longer powerful; they take their power from the airplanes or the long-range missiles that they carry. An adversary setting out to sink a ship does not send another ship against it, but instead sends an airplane or a missile.

Contrary to what the story of Billy Mitchel would suggest, the meaning of the airplane for naval warfare was perceived at an early date. In 1909 Clément Ader, a French experimenter credited by a few Gallic chauvinists with the first accomplishment of powered flight, wrote a startlingly accurate prediction of the importance of the airplane to naval warfare; and in 1911 Eugene Ely made the first takeoff from the deck of a ship—specifically, from a sloping, eighty-five-foot long platform built over the bow of a U.S. Navy cruiser. Some staff officers, their attention fastened upon the details of past wars, dismissed the primitive experiments; others saw where they would lead.

Although Ely demonstrated a landing on a ship's platform as well, progress was made more rapidly in taking off than in returning to the ship. During World War One a number of cruisers were equipped with launching platforms, and some ships carried patrol seaplanes which were hoisted to and from the ships by cranes but operated from the water. The results of naval aviation were mixed, hampered by poor equipment, inexperience, bad weather, and undependable communications. The British, who were to be pioneers in aircraft carrier development until after the Second World War, essayed a number of innovations, such as having fighters take off from a barge being towed behind a ship at high speed. One fighter thus launched shot down a Zeppelin at 20,000 feet. The fighters of those days could fly slowly, and between the sea winds and the headway of the ship they were nearly airborne while parked on the deck. Sopwith Pups demonstrated the remarkable ability to take off in a distance of twenty feet; they were even launched from platforms built out along the barrels of the

An XF9C-1 "traps" aboard the navy dirigible Macon in the early 1930s. The ill-fated airships Akron and Macon were conceived as flying aircraft carriers, but both were destroyed in storms.

heavy guns of cruisers, whose turrets could be rotated into the relative wind.

Since the early carriers were all cruisers or other types modified to launch aircraft, their large, central superstructure presented an obstacle to landing aircraft. Launches could be made from the bows of the ship, heading into the wind, in undisturbed airflow. But returning was different; it was necessary to land, as well as to launch, into the wind, and this meant approaching from behind the ship. The superstructure disturbed the airflow over the rear deck, and landings were difficult. An intrepid Englishman, Commander Dunning, demonstrated that it was possible to land a Pup on the takeoff deck as the HMS *Furious* steamed into the wind: he dipped down alongside the bow of the ship at minimum speed, sidestepped to place himself over the foredeck, and lowered his wheels to the deck while men ran up to hold him down by straps specially attached to the airplane. He succeeded twice, the Pup practically hovering over the deck in the wind; the third time the airplane went overboard. Dunning was rescued, and the *Furious* was subsequently fitted with an aft landing platform. Because of the disturbed airflow over the aft platform, however, landings were considered impractically dangerous.

The innovative English built the first flush-deck carrier, the HMS *Argus.* Nicknamed the "Flatiron," it had no protruding superstructure whatever. The subsequent HMS *Eagle* introduced the offset "island" superstructure; and in 1919 was launched the HMS *Hermes,* the first ship to have been laid down expressly as a carrier, rather than converted from some other type after launching or while in construction.

By the end of World War One, the importance of carriers was widely recognized, even though they had not played a major role in the war. The strategic relationship of air and sea power was still obscure; uncertainty about it was reflected in the political battles fought in several countries over the issue of whether all flying personnel and machines were to be subsumed into a new separate branch of the service, or whether naval aviators, for instance, would remain naval officers.

Billy Mitchell, who is famous as a visionary promoter of military aviation, actually saw no future for the aircraft carrier. His famous demonstrations of the sinking of surface ships by aircraft were intended to demonstrate not the potential of naval aviation, but the power of land-based aviation against the navy. If a ship were close enough to land to play a role in warfare, he claimed, it was close enough to be sunk by aircraft. The fact that Mitchell's aircraft did, in fact, sink several ships was inconclusive; the ships were stationary, they did not defend themselves, no effort was made to repair them, and several actually survived a great many attacks over a period of days before sinking.

The vulnerability of a battleship costing many millions of dollars to a few airplanes and bombs costing only thousands was not, as it turned out, destined to eliminate navies. But it was destined to eliminate battleships. An eighteen-inch gun was indeed meaningless against aircraft, and for that matter against the submarine, whose development only slightly preceded that of the carrier. But aviation did not make navies obsolete, as Mitchell thought it would; instead, navies were to make aviation over

A T4M biplane circles the straight-deck Saratoga in the years before World War II.

Crew abandoning the burning Lexington in the Coral Sea, 1942.

90

into a weapon of their own. The carrier would be the battleship of modern times; airplanes would be its guns.

Between wars, the United States began building carriers; at the christening of one of them, the original *Saratoga*, a fateful prophecy was uttered: " . . . a bombing attack launched from such carriers, from an unknown point, at an unknown instant, with an unknown objective, cannot be warded off by defensive aircraft based on shore."

Pearl Harbor bore out the prophecy. That raid was a titanic application of the carrier to the old naval role of shore bombardment; but now the Japanese fleet could strike from hundreds of miles away, and yet with pinpoint accuracy. The saving grace of Pearl Harbor for the United States was that, ironically, if this was the classic blow of carriers against ships and land bases, the one type of ship that was not to be found and sunk at Pearl Harbor was the one type that could answer the Japanese challenge: the carrier.

The British had lost several carriers early in the war: the *Courageous* and the *Ark Royal* to submarines, the *Glorious* to a pair of German battle cruisers (in apparent contradiction of the superiority of carriers—but circumstances were sometimes unkind), and the *Hermes* to aircraft. Conversely, British carrier aircraft engaged in highly successful raids against the French fleet at Oran and the Italian at Taranto—both turkey shoots, like Pearl Harbor, against ships immobilized in harbors.

But it was in the Pacific that the major chapters of the war history of the carrier were to be written.

The Japanese raid against Pearl Harbor began with a seven-day cruise from the Japanese home islands, swinging far northward, despite bad weather, to avoid the sea lanes and preserve the element of surprise. A chance encounter with another vessel could have ruined everything, but there was no chance encounter. While the Japanese ambassador in Washington knocked on the door of Secretary of State Cordell Hull to deliver a message terminating talks on the Pacific situation, 350 planes were launched against Pearl. At the end of the attack, 30 Japanese planes and 35 airmen had been lost; the American toll was 15 ships sunk or heavily damaged, 4 of them battleships; 188 aircraft destroyed on the ground; and 2,335 dead.

The great carrier battles of the Pacific war were battles of a new kind. That of the Coral Sea, fought on May 7–8, 1942, cost the United States one of its three remaining carriers, the *Lexington*. As she slowly foundered her men sat on the deck edge eating ice cream, waiting to be rescued. The battle of the Coral Sea set the new style in naval warfare: the ships involved never came within sight of one another. They had become like miniature nations, waging air wars on frontierless seas.

In June 1942 the Japanese attempted to seize Midway Island, the westernmost end of the archipelago at whose eastern extremity Hawaii lay. The Japanese fleet approached Midway in two segments, the faster carriers preceding the cruisers and battleships. The Japanese carriers proved amply able to defend themselves against attack from obsolescent American torpedo planes; fighters and guns repulsed and destroyed squadron after squadron as they made their long horizontal passes at the ships. But the hours of torpedo attacks

pulled the defending airplanes down low and exhausted their fuel and ammunition, and when a squadron of U. S. dive-bombers finally arrived they were unopposed and were immediately able to cripple two carriers while Japanese deck crews raced to rearm and relaunch their fighters.

The Japanese navy lost four fleet carriers in this battle; the United States only one. Midway, one of the decisive battles of military history, underscored the vital importance of timing in carrier warfare; there were times when the carrier was powerful, others when it was weak, depending on the position and readiness of its airplanes. During launches and recoveries, the carrier had to steam into the wind and was not free to maneuver. Simultaneous launches and recoveries were difficult. So the carrier had to minimize the time spent getting its airplanes refuelled, rearmed, and into the air; and from this tactical obligation grew the elaborate rituals of the deck. Haste was, and still is, everything. Like a gun being reloaded, a carrier in the process of launching or recovering its planes was useless.

Despite the slaughter of torpedo planes at Midway, aircraft were astonishingly effective against surface ships. Of thirty-nine Japanese battleships, carriers, and heavy cruisers sunk during the war, twenty-seven were sunk by torpedo planes and dive-bombers, eight by submarines, and only four by other surface ships. The battleship, on the other hand—what had once gone by the imposing name of "dreadnought"— had become nearly useless as an offensive weapon.

An F6F Hellcat readies for a rolling takeoff aboard Yorktown *in 1944.*

Shipboard fighters were also developed to a high level of effectiveness against the excellent Japanese Zero fighter. The requirement of landing on a carrier led to navy fighters having lower wing loadings, and hence better maneuverability, than was necessary in land-based fighters; and, especially early in the war, it forced upon navies an emphasis on pilot technique which has become traditional. Navy fighter pilots today still consider themselves the best dogfighters in the world.

Although the British-Argentine mini-war in the Falklands was a significant sea battle involving carriers on the British side, there have been no battles between carriers since World War II. Preeminent in the world in carriers, the United States has used them as its "big stick" in gestures of gunboat diplomacy, in the battleship's old role of shore bombardment, and as a political lever. In these roles carriers have been effective; and it is in the expectation that these roles will continue to be important that carriers continue to sail, to train, to practice, and to deploy.

The carrier's role in modern warfare is expressed in terms such as "power projection" and "showing the flag," and in the older formula of "sea control." Sea control is the traditional role of navies; little Britain was powerful, we are told by our history books, because she controlled the seas. Sea transportation is still absolutely vital to national survival because petroleum and many other vital resources travel by sea. If, for example, the Soviet Union attempts to blockade oil shipments bound here from the Persian Gulf, a carrier, as its ancestor the battleship would have done, would keep the sea lanes open, its airplane-guns ranging hundreds of miles, its antisubmarine pa-

The Carl Vinson under construction. In a turnaround from the policy of the late 1970s, the Reagan administration authorized construction of three more Nimitz-class carriers.

trols stalking vast areas of water. To some extent, sea control also implies control of neighboring land areas.

Showing the flag, on the other hand, is political in purpose. The navy is the most visible of the portable portions of our military establishment. The United States cannot cow truculent enemies by firing over their heads with ICBMs; but it can steam into the vicinity with a fleet, park a carrier in a foreign port, and scribble messages of warning in the sky with the contrails of fighters.

Vietnam was an instance of power projection. Carriers sailed up and down the Tonkin Gulf, launching strikes from one end of that narrow country to the other. Everything was in convenient reach, and the Vietnamese did not have the wherewithal to

retaliate. The carrier could wage war continually.

In recent years, power projection has been demonstrated as carrier-launched A-6s and A-7s, accompanied by their usual host of covering fighters, tankers, AWACs, and electronic-jamming aircraft, have bombed targets in Lebanon and Libya.

All-out naval warfare would be a different story; the carriers would undoubtedly be the prime targets since they represent America's, and NATO's most potent offensive striking force at sea. Questions of the big carriers' vulnerability to attack are commonplace, especially given the stupendous cost—approaching $4 billion—for one of the new nuclear ships and its aircraft. They're easy to locate with space satellites, say the detractors; what if the Soviets hit a

94

carrier with a nuclear weapon?

Such a strike would of course eliminate the carrier; no one will argue the point. But should a situation arise in which powerful adversaries are vaporizing American carriers with nuclear missiles, this country will have far more to worry about than its navy. Of more relevant concern is a middle ground—neither unthreatened peace nor cataclysmic nuclear war, but something in between.

The Asian wars in Korea and Vietnam were in between; in both conflicts our carriers delivered air strikes from offshore, and no real efforts were made to attack them. Such immunity can hardly be guaranteed forever, but the large carriers are far less vulnerable than one might expect—less vulnerable to conventional attack, goes current navy thinking, than any of our ships. The carrier's great size makes it no more detectable than any other ship, with observation satellites that can pinpoint floating objects the size of an oil drum. Its size does make it harder to destroy, however, and the current ships are heavily protected by armor, compartmentalization, automated damage control, and redundant firefighting systems and training. Several horrible fires, explosions, and deck crashes on carriers in the sixties and seventies taught the carrier forces a great deal about damage, although at a painful human cost.

Beyond its own ability to absorb damage from cruise missiles or torpedoes, the carrier protects itself, and its companion ships, in depth by launching its aircraft to perform their various missions. Fleet-defense F-14s can head off missile-carrying aircraft at ranges of several hundred miles. Its antisub aircraft can do much to "sanitize" the sub-surface world around and ahead of the task force. And its early-warning Hawkeyes can track every target in a million cubic miles of airspace. Getting through the carrier's elaborate defenses would not be easy.

Dispersal of aircraft onto smaller and cheaper carriers is a commonly mentioned alternative to continued dependence on the ultra-expensive nuclear giants. But the objection is that small ships can't handle the ideal interlocking mix of aircraft—fighters, attack bombers, tankers, early-warning birds, ELINT planes, and antisub patrol craft. Their lower cost would thus bring on decreased effectiveness and greater vulnerability to attack. The British learned this when they were forced to fight a naval and air war in the South Atlantic with only a pair of helicopter-assault carriers, roughly comparable to our Harrier-and-helicopter LHA/LPH assault ships. Without airborne early warning, they were forced to position small frigates and destroyers as radar picket ships, and several were lost or damaged. In addition, only a handful of Harriers, although magnificently flown far beyond their normal capabilities by RAF and Royal Navy pilots, were available to carry out strikes in support of ground troops fighting on the Falklands.

To the deep satisfaction of big-carrier advocates in the U.S. Navy, the Soviets have finally admitted to the most glaring weakness in their flashy new surface navy—an almost total lack of blue-water air cover. At least two conventional large-deck carriers are now known to be under construction, undoubtedly intended to field navalized versions of capable aircraft such as the MiG-23 Flogger and the new generation of Soviet fighters, the "RAMjets." At present the So-

95

Enterprise *(CVN-65) leaves San Francisco Bay. Its air wing will fly aboard when the ship hits the high seas.*

96

viets, like the British, operate only a few V/STOL carriers, which they use primarily as antisubmarine vessels. In addition to ASW helos, these ships launch the YAK-36 Forger, an unimpressive VTOL jet fighter of extremely limited capabilities.

Soviet endorsement of the large-deck concept, along with the near-catastrophic experience of Great Britain in the South Atlantic, has strengthened the hand of big-carrier advocates. Against all expectations of a decade ago, three new supercarriers have been funded. The *Theodore Roosevelt* (CVN-71) is now in the water and will soon enter active service. It will be followed by the *Abraham Lincoln* (CVN-72) and the *George Washington* (CVN-73). It is expected that the last of these ships will still be in active service in the year 2040. Thus the U.S. Navy will move toward maintenance of fourteen carrier battle groups, with a fifteenth ship always in SLEP (Service Life Extension) until 1998. The arrival of the *George Washington* will lead to the retirement of the *Coral See* (CV-43) after 45 years of continuous service.

Modern aircraft carriers are the mightiest and most elaborate extension ever attempted of the armed hand. They are the successors of the battleships, using aircraft for guns and extending their ranges thereby from tens to hundreds, even thousands, of miles. The wielder of the carrier—the captain, the admiral—does not command at a distance; he is there, aboard the ship; he surveys the battlefield with radar eyes and follows its progress on status boards, and if the ship is attacked or sunk, it is he who is attacked, he who may die.

Machines change, languages change, costumes change, but men remain the same. The aircraft carrier in many ways resembles the medieval walled town. It embodies hierarchical relationships, assumptions, rituals, psychological roles, and even physical conditions, which have not changed in hundreds or thousands of years. Only the trappings change. Carrier pilots are a modern knighthood, supported by the feudal society of the ship. Their confraternity is cemented by the conviction that they are the best pilots in the world; that conviction is nourished, in turn, by the obvious dangers of their work. Being carrier pilots provides them not only with pride, but with an unforgettable identity, a biographical landmark of which, to their deaths, long after leaving the navy, they will never lose sight.

Wars come in all styles, some of them much less up-to-date than others; for this reason, carriers have proven useful and will be for years to come. But today's giants are a pinnacle, a climax in the evolution of their species. It is improbable that in the warfare of future generations so much will

be entrusted to mere human skill, such chances routinely taken, or such difficult, elaborate, phantasmagorical means used to achieve political ends. Computers, satellites, and missiles will inherit the martial work and carry it on.

The future is as murky now as it has ever been in the past. Cataclysms of every sort loom ahead of us. Yet they seem uncertain, arbitrary, and insubstantial. We plunge toward them at an accelerating pace, the present dissolves into a remote and unimportant past. It is not hard to imagine the time when we will glance backward to see, about to vanish in the last fragment of a vaporous and shifting light, the battlements, terraces, and turrets of a castle moated by the sea; and will feel, in a sentimental memory purged of bloodshed, terror, and futile bereavements, only nostalgia for the receding shadow, for its silenced weapons, its rusting grandeur. And then the last giant carrier will be gone.

Glossary

ACLS Automatic Carrier Landing System. The autopilot uses signals from the ship's radars to guide the airplane to the deck without action by the pilot.

Air Boss Air Department Officer responsible for all flight and hangar deck operations.

Air wing Complete complement of aircraft on a carrier.

Angle of attack Angle of the wing to the flight path of the airplane; determines amount of wing lift at a given speed, and is used as a speed reference during approach.

APC Speed-holding automatic throttle.

ASW Antisubmarine Warfare.

Ball The primary optical landing aid, formally called the FLOLS, MOVLAS, or LLD.

Barrier net, barrier rig A barricade made of strips of nylon webbing used to stop an airplane whose arresting hook has malfunctioned.

Bingo field Land runway to which an airplane diverts if it cannot land on the carrier.

Blue shirt Deck hand who chocks and chains aircraft, operates elevators, drives tractors.

Blue water ops Flight operations carried on out of reach of a bingo field.

Bolt, bolter An unsuccessful landing in which the airplane fails to engage the arresting cable; distinguished from a wave-off, in which the approach is terminated before the airplane reaches the deck.

Box A metal channel used to position the nosewheel for hookup and launch.

Break out Emerge from clouds.

Brown shirt Plane captain, responsible for the care and cleanliness of an airplane.

Burner Afterburner; a device for feeding additional fuel into the exhaust of a jet engine to increase its maximum thrust.

Canard surfaces Small supplementary wings located ahead of the main wings on some aircraft.

Cat, catapult Steam-powered device used to accelerate airplanes to flying speed during launch.

CATCC Carrier Air Traffic Control Center.

CCA Carrier Controlled Approach; the pilot is "talked down" the glidepath by a radar controller on the ship. Equivalent to a GCA (Ground Controlled Approach) at a land runway.

Checkers See White shirts.

Chock Metal frame placed around the tire of an airplane to keep it from rolling.

CIC Combat Information Center.

COD Carrier Onboard Delivery; plane used to bring personnel and materiel onto the carrier.

Coupled approach An approach to a landing flown with the autopilot coupled to radio navigational aids.

Cranial Deck crewman's safety helmet.

Cross-deck pendant The portion of the arresting cable which stretches across the deck and engages the hook.

CV Navy designator for a carrier, as CV-64 (the *Constellation*). CV (N) means a nuclear-powered carrier.

Cyclic ops Alternating launches and recoveries at intervals of an hour or two.

DLC Direct Lift Control; an intermediate setting of the flight path spoilers on some aircraft, such as the F-14A and S-3.

99

Dropline A chain of orange lights hanging down the stern of the ship, on the centerline, which enables pilots to make precise adjustments in line-up.

Elevator Moveable portion of the edge of the flight deck, used for transporting aircraft between the flight and hangar decks.

ECM Electronic Countermeasures; systems for jamming enemy detection, weapons, and communication equipment.

Flare The transition from a descent to level flight just prior to touchdown on a runway; normally omitted in a carrier landing.

Flight deck The flat, exposed deck of a carrier, on which launches and recoveries take place.

FLOLS Fresnel Lens Optical Landing System; the principal optical landing aid, colloquially called the ''ball.''

Fly by wire Electronic rather than hydraulic or mechanical operation of aircraft control surfaces.

FOD Foreign Object Damage, usually the result of jet engines sucking up debris from the deck or ground.

Fouled deck Any condition in which the deck is not ready to receive a landing aircraft.

Gedunk The ship's junk-food store.

Glidepath The path of descent to a landing.

Glideslope The electronic or optical beam marking the glidepath.

Glove vanes Retractable canard surfaces in the wing root fairing, or ''glove,'' of the F-14.

Green shirt A deck hand whose duties involve readying airplanes for launch, securing them to the catapults, etc.

G-suit Flying suit that provides compressed air pressure on an air crewman's lower body to reduce the effect of G-load.

Handler Person in charge of arranging airplanes on the flight and hangar decks.

Hangar deck The main deck of the ship, on which airplanes are stored and maintained; usually four decks below the flight deck.

Holdback Hourglass-shaped breakable link which secures planes to the deck up to the moment of launch.

Hook The arresting hook which is lowered from the tail of the airplane for the landing approach, and which catches the arresting cable.

Hook aim point The point on the flight deck, between the second and third arresting cables and about a third of the length of the angled deck forward from the ramp, at which the hook should ideally strike the deck.

Hook to eye value The vertical distance between the hook and the pilot's eyes in the landing attitude.

Hot pump To fuel an airplane while its engines are running.

HUD Head Up Display; projects flight instrument indications on a transparent screen inside the windshield, so that the pilot need not look down to read the instruments.

ICBM Intercontinental Ballistic Missile.

ILS Instrument Landing System.

Island The portion of the ship's superstructure protruding above the flight deck, always on the right side.

Jet blast deflector A barrier which is raised out of the flight deck behind an airplane about to be launched, and which deflects its jet exhaust upward.

Kneeknocker Oval opening in bulkhead for a passageway.

LLD Light Landing Device; the ball.

LSO Landing Signal Officer; in charge of recoveries, judges the quality of each approach, guides and critiques pilots, gives wave-offs if necessary.

MAD Magnetic Anomaly Detection equipment, used to sense the presence of a submarine beneath the surface.

Main deck The deck from which the numbering of other decks starts; on a carrier it is the hangar deck, not the flight deck.

Mangler See Handler.

Marshall To enter holding patterns behind the ship prior to approach; also, as a noun, the holding area, and the state of being there (at marshal, in marshal).

Meatball See Ball.

Military power Maximum engine power.

Mouse Self-contained two-way radio headphones worn by some flight deck personnel.

MOVLAS Manually Operated Visual Landing System; manual backup system for the FLOLS, which is automatic.

Mule Small tractor that moves airplanes on the flight and hangar decks.

Nacelle A streamlined enclosure around an airplane engine located outside the fuselage.

Nose tow A bar that protrudes from the nosewheel strut of an aircraft, by which it is secured to the catapult.

Pickle switch Pushbutton switch by which the LSO illuminates a cluster of red lights around the ball to wave off an approaching aircraft.

Precision Approach Radar Radar that displays height as well as position of approaching aircraft.

Primary Air Controller. Same as Air Boss.

"P" school Three-week orientation course that includes flight-deck theory and first aid.

Purple shirt Deck hand in charge of fuelling aircraft.

Rack Bunkbed for enlisted men.

Red shirt Deck hand who handles bombs, ammunition, and fire equipment.

Retract engine The motor that pulls the arresting cable taut after an arrestment (pronounced REtract).

RIO Radar Intercept Officer; backseat crewman on the F-14A and the F-4.

Rounddown The aft end of the flight deck; the ramp.

Senso Sensor operator on an S-3 who interprets information from listening devices dropped by the airplane into the water.

Settle in close To drop too low during the last few hundred feet of the approach.

Shuttle The portion of the steam catapult that protrudes above the deck, and to which the nose tow is secured.

SLEP Service Life Extension Program; a program to overhaul and modernize, one by one, the existing carriers.

Snipe Ship's powerplant operator and engineer.

Spoolup Bringing a jet engine up to speed.

Spot the deck Look at the deck, rather than the ball, during an approach.

TACAN System to measure the distance and bearing between a ship and an airplane.

Tacco Tactical Coordinator; S-3 crewman who manages the attack on a submarine once it has been detected by the Senso.

Trap A successful arrested landing; opposite of bolter.

VDI Vertical Display Indicator; a cathode ray tube navigation display.

VFR Visual Flight Rules; weather conditions good enough for visual flight, and the procedures used under those conditions.

V/STOL Variously, Vertical/Short or Very Short Takeoff and Landing.

VTOL Vertical Takeoff and Landing.

Vulture's Row Open-air observation deck on the top of the island.

Waveoff Order from the LSO to the pilot telling him to abort an approach; may be due to quality of the approach or to a fouled deck.

White shirt Deck crewman who inspects airplanes coming up to the catapults for launch.

Yellow shirt Director of movement of airplanes around the flight deck.

101

Mitzi Trumbo

John Blaustein

THE AUTHOR

Peter Garrison has been a contributing editor of *Flying* magazine for the past seventeen years. He is well known in the aviation community for the design and construction of *Melmoth*, a single-engine, retractable-gear homebuilt airplane, in which he flew across the Atlantic and Pacific oceans. To experience the sensations and techniques of carrier flying, Garrison conducted his research on carriers at sea and in land-based carrier landing simulators. Garrison has written several other books: *Homebuilt Airplanes* (Chronicle Books), *Flying Airplanes: The First Hundred Hours* and *Ocean Flying* (Doubleday).

THE PHOTOGRAPHER

George Hall is a San Francisco photographer specializing in aerial and aviation photography. His photographs of carrier flying were taken on carriers in both the Atlantic and Pacific oceans. To obtain action shots he photographed on deck, from helicopters, and from planes on cat take-offs and in arrested landings. Hall's photographs illustrate other Presidio AIRPOWER books—*USAFE: A Primer of Modern Air Combat in Europe, RED FLAG: Air Combat for the '80s, MARINE AIR: First to Fight*, and *TOP GUN: The Navy Fighter Weapons School*—and also are featured in the annual AIRPOWER Calendar.

THE PRESIDIO POWER SERIES presents exciting, graphic profiles of all aspects of the modern American military—on air, sea, and land. Each book is based on first-hand interviews and personal experience, and is profusely illustrated with action photographs. Each *POWER SERIES* book takes you inside and makes you a part of the world's most potent fighting forces.

They are published in a uniform $8 \times 8\frac{3}{4}$ softcover format.

AIRPOWER BOOKS:

#1001 CV: Carrier Aviation (New Revised Edition)
Text by Peter Garrison and George Hall
Photography by George Hall
Available Fall 1987   ISBN: 0-89141-299-9

#1002 USAFE: A Primer of Modern Air Combat in Europe
Michael Skinner
Photography by George Hall
$10.95   ISBN: 0-89141-151-8

#1003 RED FLAG: Air Combat for the '80s
Michael Skinner
Photography by George Hall
$12.95   ISBN: 0-89141-168-2

#1004 SAC: A Primer of Modern Strategic Air Power
Bill Yenne
$10.95   ISBN: 0-89141-189-5

#1005 MARINE AIR: First to Fight
John Trotti
Photography by George Hall
$12.95   ISBN: 0-89141-190-9

#1006 SPACE SHUTTLE: A Quantum Leap
George J. Torres
$12.95   ISBN: 0-89141-253-0

#1007 TOP GUN: The Navy's Fighter Weapons School
Text and photographs by George Hall
$12.95   ISBN: 0-89141-261-1

#1008 STRIKE UNIVERSITY: The Navy's Attack School
John Joss
Photography by George Hall
Available Fall 1988   ISBN: 0-89141-320-0

SEAPOWER BOOKS:

#2001 USN: Naval Operations in the '80s
Michael Skinner
$12.95   ISBN: 0-89141-209-3

#2002 USCG: Always Ready
Text and photographs by Hans Halberstadt
$12.95   ISBN: 0-89141-256-5

LANDPOWER BOOKS:

#3001 AIRBORNE: Assault from the Sky
Text and photographs by Hans Halberstadt
Available Fall 1987   ISBN: 0-89141-279-4

#3002 GREEN BERETS: U.S. Army Special Forces
Text and photographs by Hans Halberstadt
Available Spring 1988   ISBN: 0-89141-280-8

#3003 USAREUR: U.S. Army in Europe
Michael Skinner
Available Fall 1988   ISBN: 0-89141-311-1

#3004 RANGERS: We Lead the Way
Text and photographs by Hans Halberstadt
Available Spring 1989   ISBN: 0-89141-318-9